EVOLUTION 101

EVOLUTION 101

RANDY MOORE AND JANICE MOORE

SCIENCE 101

GREENWOOD PRESS
Westport, Connecticut • London

Library of Congress Cataloging-in-Publication Data

Moore, Randy.
    Evolution 101 / Randy Moore and Janice Moore.
       p.   cm. — (Science 101)
    Includes bibliographical references and index.
    ISBN 0–313–33292–4 (alk. paper)
    1. Evolution (Biology)—Popular works.   I. Moore, Randy.   II. Title.
  QH367.M81   2006
  576.8–dc22   2006020537

British Library Cataloguing in Publication Data is available.

Library of Congress Catalog Card Number: 2006020537
ISBN: 0–313–33292–4
ISSN: 1931–3950

First published in 2006

Greenwood Press, 88 Post Road West, Westport, CT 06881
An imprint of Greenwood Publishing Group, Inc.
www.greenwood.com

Printed in the United States of America

The paper used in this book complies with the
Permanent Paper Standard issued by the National
Information Standards Organization (Z39.48–1984).

10 9 8 7 6 5 4 3 2 1

To Mom and Dad,
With love and gratitude that go far beyond words.

CONTENTS

SERIES FOREWORD

What should you know about science? Because science is so central to life in the 21st century, science educators believe that it is essential that *everyone* understand the basic foundations of the most vital and far-reaching scientific disciplines. *Science 101* helps you reach that goal—this series provides readers of all abilities with an accessible summary of the ideas, people, and impacts of major fields of scientific research. The volumes in the series provide readers—whether students new to the science or just interested members of the lay public—with the essentials of a science using a minimum of jargon and mathematics. In each volume, more complicated ideas build upon simpler ones, and concepts are discussed in short, concise segments that make them more easily understood. In addition, each volume provides an easy-to-use glossary and an annotated bibliography of the most useful and accessible print and electronic resources that are currently available.

PREFACE

Evolution 101 is a book about evolution. As the title suggests, *Evolution 101* is not meant to be a comprehensive treatment of evolution; even if we tried to do that, we could not, for the topic is far too broad. Instead, we have focused on the core concepts of evolution—the history of evolutionary thought, the evidence for evolution, how evolution works, the scale and products of evolution, and evolution in our daily lives. By the time you finish this book, you should have a good understanding of what evolution is, why it is important, and how it continues to help us understand a variety of aspects of life. In an effort to make the book accessible to a broad range of people, we have kept references to other scientific fields at a minimum. We hope that this book will prove useful to readers even if they have had limited exposure to biological education.

In 1973, Nobel Laureate François Jacob noted that "there are many generalizations in biology, but precious few theories. Among these, the theory of evolution is by far the most important, because it draws together from the most varied sources a mass of observations which would otherwise remain isolated; it unites all the disciplines concerned with living beings; it establishes order among the extraordinary variety of organisms and closely binds them to the rest of the earth; in short, it provides a causal explanation of the living world and its heterogeneity." Jacob was right; evolution *does* unify the study of life. Nevertheless, for some people, evolution is a controversial topic that raises several social and theological concerns. Although some of these issues—for example, extinctions and the age of the Earth—cannot be separated from evolution, we do not focus on them in *Evolution 101*. To be sure, these topics are nontrivial; we do not suggest otherwise. However, they are simply beyond the scope and intent of this book. Meanwhile, we believe that the ability to appreciate these and other evolution-related issues requires an

xiv   **Preface**

understanding of what evolution is, as well as what it is not. Providing general readers with that understanding of evolution is the goal of this book.

We're grateful to several people for their help with this book. Sarah Bevins provided valuable criticisms, other colleagues gave us helpful suggestions, and Alton Biggs, Kevin Downing, Jeff Dixon, Annia Fayon, and officials at Westminster Abbey helped us with some of the photographs. These and other contributions have improved the book.

We hope you enjoy learning about one of the most pivotal and powerful ideas in the history of human thought—evolution.

Randy Moore
Janice Moore

1

THE HISTORY OF
EVOLUTIONARY THOUGHT

Throughout history, people have wondered about the astonishing diversity of life, with its adaptations for survival and complex machinery. Every culture has its stories of how the world came to be, and many of the stories include amazing creations of plants and animals, imaginative dialogs among these organisms in some cases, and above all, lessons to be learned. Some people have also wondered if there could be an explanation for life's diversity that involves more routine occurrences, without supernatural events.

In this chapter, you will learn about the history of evolutionary thought. We start with the history of this idea because such an approach is an effective way of understanding the idea. As biologist Ernst Mayr noted, "Most scientific problems are far better understood by studying their history than their logic."

EARLY IDEAS ABOUT LIFE'S DIVERSITY

As early as 550 B.C., the Greek philosopher Anaximander of Miletus (610–546 B.C.) argued that Earth had not been created abruptly, but instead that life had started as slime in the oceans, later moved onto land, and that humans and all other vertebrates had descended from fish. Several decades later, the Greek philosopher Empedocles (492–432 B.C.) speculated that plants arose first after Earth formed, that animal life "budded off" from plants, and that the universe and everything in it is gradually changing (e.g., "Many races of living creatures must have been unable to continue their breed"). In contrast, around 350 B.C., Aristotle (384–322 B.C.) claimed that all species are static (i.e., do not evolve) and can be arranged in a hierarchy according to their degree of perfection, with the most perfect organisms (those would be we humans) at the top. In Aristotle's perfect universe, there were "higher" and "lower" organisms; for example, dogs were clearly "higher" than invertebrates. Every species had a clearly defined position; species did not, and could not, change. Aristotle's ideas about life became known as the Great Chain of Being, which expressed God's perfect design, wisdom, and power. The chain also inferred the method of creation, for it started with simpler forms of life and extended link by link to humans. The Great Chain of Being was described more than 2,000 years later by poet Alexander Pope in his *Essay on Man*:

> Vast chain of being! Which from God began,
> Natures ethereal, human, angel, man,
> Beast, bird, fish, insect, what no eye can see,
> No glass can reach; from Infinite to thee,
> From thee to nothing.—On superior pow'rs
> Were we to press, inferior might on ours;
> Or in the full creation leave a void,
> Where, one step broken, the great scale's destroy'd;
> From Nature's chain whatever link you strike,
> Tenth or ten thousandth, breaks the chain alike.

Aristotle's ideas were popular among Western thinkers and went unchallenged for centuries (see Chapter 4). Near the end of the 17th century, however, Western thinking began to change. It was the beginning of a sparkling, perhaps even inflammatory, era in human ideas, one that bore fruit in everything from revolutions and the spread of democratic ideals to the birth of science as we know it. Astronomy, physics, chemistry—all would change, and with them, biology.

In 1686, botanist John Ray (1627–1705) proposed the first definition of a species, and argued that these species (he classified more than 18,000 species of plants) had not changed over time. As Ray noted, "the works created by God at first" had been "by Him conserved to this day in the same state and condition in which they were first made." Later, Ray—the "father of natural history"—claimed in his *Wisdom of God as Manifested in the Works of Creation* that adaptations were permanent traits designed by God. According to Ray, nature could preserve, but not produce, a species. However, Ray suspected that the presence of fossil fishes atop mountains suggested that the mountains had been lifted over long periods of time. In 1663, after discovering a buried forest found "in places which 500 years ago were sea," he wrote, "Many years before all records of antiquity these places were part of the firm land and covered with wood; afterwards being overwhelmed by the violence of the sea they continued so long under water till the rivers brought down earth and mud enough to cover the trees, fill up these shallows and restore them to firm land again ... that of old time the bottom of the sea lay so deep and that that hundred foot thickness of earth arose from the sediment of those great rivers which there emptied themselves into the sea ... is a strange thing considering the novity of the world, the age whereof, according to the usual account, is not yet 5500 years." Ray, like others, struggled to understand the enormity of geologic time.

Ray, more than anyone else, made the study of botany and zoology a scientific pursuit. Before Ray, plants and animals had been "classified" in alphabetical order of their names. Ray changed this when he invented an orderly taxonomic system based on anatomy and physiology, thereby paving the way for history's greatest and most famous taxonomist: Swedish naturalist Carl von Linné (1707–1778; Figure 1.1). In 1735, von Linné—writing under his Latin name Carolus Linnaeus—tried to classify all life on Earth in hopes of discovering the pattern of the creation. Linnaeus' system included kingdoms, classes, orders, genera, and species, each of which was a distinct "archetype" that reflected God's omnipotence. Linnaeus also proposed that new species within genera could be produced by hybridization (i.e., by mating between species; see Chapter 4), but this hybridization was guided

Figure 1.1 Carolus Linnaeus. Linnaeus developed a system of naming, ranking, and classifying organisms that was meant to reveal the divine order of life. His system influenced generations of biologists and is still used today. (*Library of Congress, Prints & Photographs Division, LC-USZ62-11324*)

by God. Although Linnaeus was the first person to include "man" (as naturalists referred to humans in those days) in a biological classification system, he was uncertain about whether there should be a separate genus for *Homo*. As he wrote in 1746, "the fact is that as a natural historian I have yet to find any characteristics which enable man to be distinguished on scientific principles from an ape." In a letter to a colleague the next year, Linnaeus again stated his uncertainty about separating humans from apes: "I ask you and the whole world for a generic difference between man and ape which conforms to the principles of natural history. I certainly know of none . . . If I were to call man ape or vice versa, I should bring down all the theologians on my head. But perhaps I should still do it according to the rules of science."

Like many Westerners in their day and since, Ray and Linnaeus believed that God created life in the Garden of Eden. (Linnaeus' own motto was *Deus creavit, Linnaeus disposuit*—"God created, Linnaeus arranged.") Although Linnaeus accepted the biblical account of the Flood, he reasoned that such a short-lived event could not have moved living things far inland and covered them in sediment in the time available. As he noted, "He who attributes all this to the Flood, which suddenly came and as suddenly passed, is verily a stranger to science and himself blind, seeing only through the eyes of others, as far as he sees anything at all."

Linnaeus believed that there was a divine order to all organisms, and he developed his now-famous binomial (two-name) system of classification to reveal this order. Today, scientists continue to use a modified version of Linnaeus' system of nomenclature to name species. For example, humans are *Homo sapiens*, in which *Homo* ("human") represents our genus and *sapiens* ("wise") our species epithet (see Naming Life).

NAMING LIFE

Before Linnaeus, it was customary to use descriptive Latin phrases to name plants and animals. The first word of the phrase constituted the group to which the organism belonged. For example, all known roses were given names beginning with *Rosa*, and mints were given names beginning with *Mentha*. The complete name for peppermint was *Mentha floribus capitatus, foliis lanceolatis serratis subpetiolatis*, which means "Mint with flowers in a head, leaves lance-shaped, saw-toothed, with very short petioles." And if that wasn't bad enough, the same organism often had different names. For example, wild briar rose was referred to as *Rosa sylvestris inodora sea canina* and *Rosa sylvestris alba cum rubose folio glabro*. Although such more-than-a-mouthful names were specific, they were far too cumbersome to be useful.

Linnaeus, who once commented that he could not "understand anything that is not systematically ordered," changed this by describing each organism with a binomial (the two-name system). Organisms with the most similarities were placed in the same group, called a genus, and the genus was the first word in the binomial. Similarly, each member of a genus was called a species, and the specific epithet is the second word of the binomial. Peppermint, for example, became *Mentha piperita*. Similar genera (the plural of genus) were grouped into families, similar families into orders, similar orders into classes, similar classes into phyla, similar phyla into kingdoms, and kingdoms into domains. Here are examples of how some common organisms are classified:

	Human	Chimpanzee	Wolf	Fruit fly	Sunflower
Domain	Eukarya	Eukarya	Eukarya	Eukarya	Eukarya
Kingdom	Animalia	Animalia	Animalia	Animalia	Plantae
Phylum	Chordata	Chordata	Chordata	Arthropoda	Anthophyta
Class	Mammalia	Mammalia	Mammalia	Insecta	Dicotyledoneae
Order	Primates	Primates	Carnivora	Diptera	Asterales
Family	Hominidae	Pongidae	Canidae	Drosophilidae	Asteraceae
Genus	*Homo*	*Pan*	*Canis*	*Drosophila*	*Helianthus*
Specific epithet	*sapiens*	*troglodytes*	*lupus*	*melanogaster*	*annuus*

Today's classification systems, which continue to use binomials, try to reflect evolutionary relatedness. Such systems are said to be phylogenetic. You'll learn about these systems in Chapter 4.

In 1749, French zoologist Georges-Louis Leclerc (better known by his title Count Buffon) developed the modern definition of a biological species—namely, a group of organisms that can interbreed and produce fertile offspring. Like most scientists before him, Buffon (1707–1788) believed that all organisms were created by a deity, and, like Aristotle and other predecessors, placed humans atop his hierarchical arrangement of life. However, Buffon speculated that life evolves according to natural laws, that all life has descended from a common ancestor, and that humans are related to apes. Here's what Buffon wrote in 1753: "If we once admit that there are families of plants and animals, so that the ass may be of the family of the horse, and that the one may only differ from the other through degeneration from a common ancestor, we might be driven to admit that the ape is of the family of man, that he is but a degenerate man, and that he and man have had a common ancestor, even as the ass and horse have had. It would follow then that every family, whether animal or vegetable, had sprung from a single stock, which after a succession of generations, had become higher in the case of some of its descendants and lower in that of others."

Buffon also presented a clear argument against living creatures having been individually designed by an intelligent Creator (although he did not state this in so many words). A pig, for example, "does not appear to have been formed upon an original, special, and perfect plan, since it is a compound of other animals; it has evidently useless parts, or rather parts of which it cannot make any sense, toes all the bones of which are perfectly formed, and which, nevertheless, are of no service to it. Nature is far from subjecting herself to final causes in the formation of these creatures." (You'll learn more about these "useless parts" in Chapter 2.) In 1778, Buffon also used experiments that involved heating and cooling iron balls to extrapolate that Earth was 75,000 years old.

Buffon—one of the first popularizers of science who was regarded as a major literary figure for his writing style—believed that the physical environment (and not biotic factors, such as competition) caused species to evolve, but he never proposed a convincing mechanism for how this could occur. Buffon's claims were controversial; after all, if species had been created perfect, any of the changes that Buffon proposed would have changed the species, and therefore made it imperfect. Under pressure, Buffon later recanted his claims about evolution ("I abandon everything in my book . . . contrary to the narrative of Moses"). Nevertheless, his ideas prompted some people to begin considering new ideas for the origin of life's diversity.

In the 1790s, famed French naturalist and aristocrat Georges Cuvier (1769–1832; Figure 1.2) picked up where Buffon had left off in the 1780s. Cuvier, who established the discipline of vertebrate paleontology, used fossils to prove that extinctions have occurred. This was important, for such extinctions "broke" the Great Chain of Being, in which each species had its own ideal structure and function. Cuvier devised a classification system that included four basic groups: vertebrates (animals with a backbone), mollusks (shelled animals), articulates (such as insects and segmented worms), and radiates (symmetrical animals such as starfish). Cuvier's system is no longer used, but his new, nonlinear way of classifying life was an important break with the hierarchical thinking of the past, and pointed the way ahead.

Cuvier, the first person to identify and name the pterodactyl, believed that an animal's function determined its form (his so-called "correlation of parts") and that animals' similarities resulted from common functions, not common ancestry. That is, Cuvier believed that an animal's anatomy is too interdependent to have evolved piecemeal. Cuvier rejected Linneaus' version of life's divine plan, and argued that extinctions resulted from God-driven catastrophes (he called them "revolutions") such as floods and earthquakes. After each catastrophe, Earth was repopulated by organisms that immigrated from parts of the world which had not yet been explored. Cuvier rarely mentioned the creation of new species, preferring instead to refer to their "coming into existence." Cuvier claimed that it was even possible to see where life began; as he wrote in 1825, "What is even more surprising is that life itself has

Figure 1.2 Georges Cuvier documented extinctions and established the discipline of vertebrate paleontology. Although his discoveries strongly supported Darwin's theory of evolution by natural selection, Cuvier vehemently rejected evolution throughout his life. (*Library of Congress, Prints & Photographs Division, LC-USZ62-134030*)

not always existed on the globe, and that it is easy for the observer to recognize the precise point where it has first left traces."

Although Cuvier's research provided strong evidence for evolution (you'll learn about this in the next chapter), Cuvier rejected all types of evolution and claimed that evolution was "contrary to moral law, to the Bible, and to the progress of natural science itself." No human fossils were known to Cuvier, who concluded that humans came into existence after the biblical flood.

Evolution was next heralded by the flamboyant Erasmus Darwin (1731–1802), an English physician, poet, naturalist, and grandfather of Charles Darwin. In 1794, Erasmus published one of the first theories of evolution in his book *Zoonomia, Or, the Laws of Organic Life*. Erasmus often wrote in rhymed couplets, as shown in these lines from *The Temple of Nature* (published in 1803) that describe a gradual progress of life toward higher levels of complexity and greater mental powers:

> Organic Life beneath the shoreless waves
> Was born and nurs'd in Ocean's pearly caves.
> First forms minute, unseen by spheric glass,
> Move on the mud, or pierce the watery mass;
> These, as successive generations bloom,
> New powers acquire, and larger limbs assume;
> Whence countless groups of vegetation spring,
> And breathing realms of fin, and feet, and wing.

> Thus the tall Oak, the giant of the wood,
> Which bears Britannia's thunders on the flood;
> The Whale, unmeasured monster of the main,
> The lordly Lion, monarch of the plain,
> The Eagle soaring in the realms of air,
> Whose eye undazzled drinks the solar glare,
> Imperious man, who rules the bestial crowd,
> Of language, reason, and reflection proud,
> With brow erect who scorns this earthly sod,
> And styles himself the image of his God;
> Arose from rudiments of form and sense,
> An embryon point, or microscopic ens!

Erasmus Darwin never thought of natural selection, but he did suggest that all species had originated from the same ancestor; as he noted, "Shall we conjecture that one and the same kind of living filament is and has been the cause of all organic life? Would it be too bold to imagine, that in the great length of time since the earth began to exist, perhaps millions of ages before the commencement of the history of mankind, would it be too bold to imagine, that all warm-blooded animals have arisen from one living filament?" Erasmus Darwin also suggested that species' survival was governed by "laws of nature" rather than divine authority, and that new species arose because of competition and sexual selection. However, like others before him, Erasmus Darwin had little evidence to support his claims, his arguments were not convincing, and he could not explain adaptations. Although Charles Darwin never met his eccentric grandfather (Erasmus died 7 years before Charles was born) and was not overly impressed by his scientific claims, Charles nonetheless admired his grandfather. Indeed, when Charles was 70 years old he wrote Erasmus' biography, *The Life of Erasmus Darwin*.

One of the next attempts to explain life's diversity was by Scottish farmer James Hutton (1726–1797), who in 1785 stated publicly what scientists now know: Earth is much, much older than 6,000 years (Appendix 1). A decade later, in his two-volume *Theory of Earth*, Hutton made a variety of now-confirmed observations, including the fact that Earth's molten core is pushed up to the Earth's surface (e.g., by volcanic eruption), an observation supported by his descriptions of strata of sedimentary rocks in the Scottish Highlands penetrated by veins of igneous rock such as granite (sedimentary rock consists of particles compressed beneath a body of water, and igneous rock consists of crystals formed by extreme temperatures from molten rock). Hutton, who viewed Earth as a "beautiful machine," argued that Earth's surface is continually changed by endless cycles of uplift, erosion, and sedimentation ("rest exists not anywhere"), and that Earth's geological features are produced uniformly and gradually by everyday events such as rain, wind, earthquakes, volcanic eruptions, and running water ("In examining things present we have data from which to reason with regard to what has been"). Hutton argued that this was an endless process of gradual destruction and renewal so vast that the geological record left "no vestige of a beginning, no prospect for an end."

Hutton's *Theory of Earth* was written while Hutton was ill; it is very long (2,138 pages) and difficult to read, and therefore did not attract much attention. However, Hutton's ideas were popularized in 1802 by his friend John Playfair's *Illustrations of the Huttonian Theory of the Earth*. In *Illustrations*, Playfair featured Hutton's argument that all parts of the Earth, including its inhabitants, have changed over long periods of time.

Hutton's arguments became known as *uniformitarianism* because they claimed that today's geologic processes have acted in much the same way—that is, uniformly—through time. Although Hutton did not state an exact age for the Earth, his conclusion was clear: "We find in natural history monuments which prove that those animals had long existed; and we thus procure a measure for the computation of a period of time extremely remote, though far from having been precisely ascertained." This did not mean that there *was* no beginning. On the contrary, it simply meant that any evidence for such a calculation was buried in the rocks. Although many people rejected Hutton's claims because they contradicted claims about the biblical age of the Earth, Hutton's conclusions were well supported. Today, Hutton's work is widely recognized for establishing geology's most transforming contribution to human knowledge: deep time.

Buried in the verbose pages of Hutton's *Investigation of the Principles of Knowledge* is this: "... if an organized body is not in the situation and circumstances best adapted to its sustenance and propagation, then, in conceiving an indefinite variety among the individuals of that species, we must be assured, that, on the one hand, those which depart most from the best adapted constitution, will be the most liable to perish, while, on the other hand, those ... which most approach to the best constitution for the present circumstances will be best adapted to continue, in preserving themselves and multiplying the individuals of their race ... If those organized bodies shall thus multiply, in varying conditions according to particular circumstances ... we might expect to see ... a variety in the species of things which we might term a race." Hutton supported his suggestion by citing his studies of animal breeding, noting that dogs survived by "swiftness of foot and quickness of sight ... the most defective in respect of those necessary qualities, would be the most subject to perish, and that those who employed them in greatest perfection ... would be those who would remain, to preserve themselves, and to continue the race." Similarly, if an acute sense of smell was "more necessary to the sustenance of the animal ... the same principle [would] change the qualities of the animal, and ... produce a race of well scented hounds, instead of those who catch their prey by swiftness." The same "principle of variation" would influence "every species of plant, whether growing in a forest or a meadow." However, Hutton provided no data to support his suggestion, and his idea about life's diversity faded away.

LAMARCK PROPOSES THE FIRST TESTABLE THEORY OF EVOLUTION

In 1809, French naturalist Jean-Baptiste Lamarck (1774–1829)—a protégé of Buffon and botanist to King Louis XVI—shocked Cuvier and the rest of Europe by declaring in his book *Philosphie Zoologique* that the fixity of life was an illusion. Whereas contemporaries such as Buffon had hinted at evolution, Lamarck was its champion: "... species have only a limited or temporary constancy in their characters ... there is no species which is absolutely constant." Lamarck's goal was not to only produce "a convenient list for consulting, but ... more particularly to have order in that list which represents as nearly as possible the actual order followed by nature in the production of animals: an order conspicuously indicated by the affinities which she has set between them. Thus, to obtain a knowledge of the true causes of that great diversity of shapes and habits found in the various known animals, we must reflect that the infinitely diversified but slowly changing environment in which the animals of each race have successively been placed has involved each of them in new needs and corresponding alterations in their habits." Lamarck knew that his idea would encounter resistance: "There is one strong reason that prevents us from recognizing the successive changes by which known animals have been diversified and been brought to the condition in which we observe them; it is this, that we can never witness these changes. Since we see only the finished work and never see it in course of execution, we are naturally prone to believe that things have always been as we see them rather than that they have gradually developed." The confident Lamarck then noted that "This is a truth which, once recognized, cannot be disputed."

Lamarck, who coined the terms *biology* and *invertebrate*, worked at the Museum of Natural History in Paris. He was perhaps the world's premier invertebrate zoologist, and he argued that organisms contain a "nervous fluid" that enables them to adapt to their local environments. Lamarck believed that organisms evolve to become more complex over time, that these purposeful changes are brought about by the use or disuse of acquired traits, and that these changes made the organisms better able to survive in new environments and conditions. Simply put, Lamarck believed that what an animal did during its lifetime was passed on to its offspring.

According to Lamarck, a change in the environment would change the needs of organisms living in that environment. As Lamarck noted, "Since every species has to exist in perfect harmony with its surrounding and since this surrounding is constantly changing, the species itself, too, has to change constantly, if it is to stay in a harmonic balance with its surrounding. If it would not adjust, the species would be threatened by extinction." In turn, the organisms' altered needs would change the organisms' behaviors, and these altered behaviors would lead to the greater or lesser use of different structures. The more an organism used a part of its body, the more developed that part would become (similarly, the disuse of a part would result in its decay). Lamarck referred to this idea—namely, that the use or disuse of a structure would cause the structure to enlarge or shrink—as his "First Law." This was followed by

Lamarck's "Second Law," which proposed that the changes acquired as a result of the First Law would be inherited by the organism's offspring. As a result, species would gradually change as they became adapted to their environments. For example, Lamarck argued that wading birds evolved long legs as they stretched to keep high and dry, and that giraffes evolved long necks as they stretched their necks to reach leaves high in trees. When they stretched, Lamarck claimed, their "nervous fluid" would flow into their necks and, over successive generations, cause their necks to grow longer. This need-based result (i.e., necks getting longer in order to get food) would give giraffes permanently longer necks, and these long necks would be passed to the giraffes' offspring. As Lamarck wrote, "We know that [a giraffe], the tallest of mammals, dwells in the interior of Africa, in places where the soil, almost always arid and without herbage, obliges it to browse on trees and to strain itself continuously to reach them. This habit, sustained for long, has had the result in all members of its race that the forelegs have grown longer than the hind legs and that its neck has become so stretched, that the giraffe, without standing on its hind legs, lifts its head to the height of six meters." Lamarck's idea, which came to be known as "inheritance of acquired characteristics," suggested that there was a drive toward perfection and complexity, analogous to species climbing a ladder. Lamarck did not believe that species could become extinct; instead, he believed that they evolved into different, more nearly perfect species, in what he called "transmutation."

Lamarck was his era's most renowned advocate of evolution, and his model for evolution—namely, that traits acquired by an individual's efforts are passed to offspring—was the first testable hypothesis to explain how a species could change over time. However, it was rejected (and sometimes ridiculed) by the leading scientists of his time (i.e., Buffon and Cuvier), and was later dismissed by other scientists. After all, many people acquire traits—for example, scars or lost fingers from accidents—but the offspring of these people are not born with the same scars or missing fingers. Similarly, weightlifters with big muscles do not necessarily have offspring with big muscles. However, Lamarck's idea was very popular with the public—so much so that Charles Darwin included specific allusions and references to it in later editions of his *On the Origin of Species*.

Although Lamarck's name is most often associated with his discredited "inheritance of acquired traits," Charles Darwin and many other scientists acknowledged him as an important zoologist and one of evolution's early thinkers. As Darwin noted in 1861, "Lamarck was the first man whose conclusions on the subject excited much attention. This justly celebrated naturalist ... first did the eminent service of arousing attention to the probability of all changes in the organic, as well as in the inorganic world, being the result of law, and not of miraculous interposition." Despite his great contributions to biology, Lamarck died a blind man, living in poverty and obscurity, and was buried in a rented grave. Five years later, his remains were removed and taken to an unknown location.

In 1844, Edinburgh writer and amateur geologist Robert Chambers (1802–1871) anonymously published a grandly titled book, *Vestiges of the Natural History of Creation*, which argued that the physical world could be best understood by the use of science and natural law rather than by appealing to supernatural phenomena. Specifically, Chambers argued that (1) evolution occurs via unknown laws and in a steadily upward progression, and (2) Earth was not specifically created by God, but instead by laws that expressed God's will. As Chambers noted, "How can we suppose that the august Being ... was to interfere personally and on every occasion when a new shellfish or reptile was ushered into existence? ... The idea is too ridiculous for a moment to be entertained." Chambers' book became a *cause célèbre* and was a bestseller (it sold far more copies than any of Darwin's books) that attracted much attention; it was in its tenth edition a decade later, and was read by people ranging from Abraham Lincoln and Queen Victoria to Alfred Tennyson. However, many religious leaders condemned the book. Scientists did, too, because Chambers' science was shoddy. Although some scientists in the early 1800s worried that "Mr. Vestiges" would cause the intellectual public to reject subsequent ideas about evolution, Chambers' book showed scientists the obstacles they would have to overcome if they were to convince others of the validity of their ideas. In the early 1800s, such opposition was a moot point, for there were no convincing and testable models to explain life's diversity. Charles Darwin would change that.

CHARLES DARWIN

Charles Robert Darwin (1809–1882; Figure 1.3) was born in Shrewsbury, Shropshire, England (about 160 miles northwest of London), on February 12, 1809, into a wealthy, well-known family. Charles' father, Robert Waring Darwin (1766–1848), was a doctor (like his father, Erasmus). His mother, Susannah Wedgwood Darwin (1765–1817), was the daughter of Josiah Wedgwood, one of the founders of the Wedgwood pottery company known for its "blue china" and opposition to slavery.

Figure 1.3 Charles Robert Darwin (1809–1882) in his old age. Darwin's monumental *On the Origin of Species*, which was published in 1859, changed the course of biology. (*Image copyright History of Science Collections, University of Oklahoma Libraries*)

(Josiah Wedgwood & Sons, Limited, remains a thriving company today.) Charles had one brother (Erasmus, named after his eccentric grandfather), three older sisters (Marianne, Caroline, and Susanne), and one younger sister, Emily Catherine. The young Charles Darwin attended Shrewsbury Grammar School, but much preferred hunting and collecting shells, coins, and rocks to learning Greek and Latin. In the autumn of 1825, Charles enrolled in Edinburgh University to study medicine. There, he learned taxidermy (which would be very useful for him later), but again he didn't like his classes, and dropped out. He would not continue the family tradition of medicine.

In 1827, Charles enrolled at Cambridge University and began studying to be a clergyman. However, he continued to prefer the outdoors. He became especially interested in collecting insects, and it was at Cambridge that Charles began to appreciate the vast diversity of species, an appreciation that is inescapable if insects are the focus. While at Cambridge, Charles was mentored by Robert Grant, an admirer of Charles' grandfather Erasmus and a supporter of Lamarck's idea about inheritance of acquired characteristics. Darwin also learned about natural theology, which sought to understand God by studying God's creation (i.e., nature). To advocates of natural theology, God's goodness was visible in the way He had created a progression of life from "lower" to "higher" forms, culminating in the special creation of humans.

The leading proponent of natural theology was Anglican Archdeacon William Paley (1743–1865), who believed that evolution was unnecessary and unthinkable. Paley's book, *Natural Theology: or, Evidences of the Existence and Attributes of the Deity, Collected from the Appearances of Nature,* was a standard textbook for theology students. Paley argued that God's existence and attributes are manifest in the design evident in nature, such as in the hinges of bivalve shells and the wing-like structures that help some plants disperse seeds. Paley focused on a seductive analogy: "In crossing a heath, suppose I pitched my foot against a stone and were asked how the stone came to be there." Perhaps, for all Paley knew, it had been there forever. "But suppose I had found a watch upon the ground, and it should be inquired how the watch happened to be in that place." Paley argued that the watch's complexity was evidence of design, and that design required a designer. He then made a now-famous analogy: If a complex structure such as a watch has a designer, then a more complex structure such as an eye must also have a designer (i.e., God). As Paley noted, "Design must have had a designer. That designer must have been a person. That person is GOD."

While at Cambridge preparing to join the clergy, Darwin studied and enjoyed Paley's book; as he noted, "I do not think I hardly ever admired a book more than Paley's *Natural Theology*. I could almost formerly have said it by heart." Paley's book later became a favorite of creationists promoting "intelligent design" as a way of weakening the teaching of evolution in public schools (see Appendix 2).

While at Cambridge, Darwin lived in the same dormitory rooms that had housed William Paley 70 years earlier. Darwin studied to be a Church of England cleric; he had traditional religious beliefs, but a noticeable lack of religious zeal. Although Darwin later claimed that some of his time at Cambridge "was sadly wasted," in 1831 he graduated at age 23 and earned a Bachelor's Degree in Theology, ranking tenth among non-honors students. However, Darwin was drawn to the prospect of travel—it was a great age of exploration—and he hoped to delay his entry into religious work until he had visited more exotic locales.

Darwin Sets Sail Aboard the *Beagle* and Sees the World

In the 1830s, the British government commissioned the 90-foot ship HMS *Beagle* to set forth on expeditions "devoted to the noblest purpose, the acquisition of knowledge." Specifically, the crew of the *Beagle* was to test new clocks for the British Navy, and to map and search the coasts and harbors of South America and elsewhere for new resources. The *Beagle*'s captain was the temperamental 26-year-old Robert FitzRoy (1805–1865), and FitzRoy needed a naturalist for the voyage. FitzRoy (who would

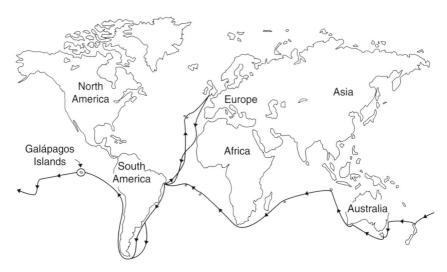

Figure 1.4 The *Beagle* sailed more than 40,000 miles between its departure on December 27, 1831, and its return to England on October 2, 1836. During this time, Darwin spent 18 months at sea; the remainder of the time was spent on land investigating local life and geology. The *Beagle* was captained by Robert FitzRoy, who, like Darwin, was later elected a Fellow of the Royal Society. FitzRoy, one of the first advocates of weather forecasting, rejected Darwin's theory of evolution; he felt guilty that his voyage was used to undermine the Bible. (*Jeff Dixon*)

later reject Darwin's theory) wanted the naturalist to be cleric-botanist John Henslow (1796–1861), a friend of Darwin's. (Darwin had attended Henslow's botany lectures at Cambridge.) Henslow declined FitzRoy's offer, but recommended Darwin for the job. Charles' father did not want him to go on the voyage (ships such as the *Beagle* were often called "floating coffins"), but Josiah Wedgwood (Charles' uncle) convinced Charles' father to let his son go on the cruise. Darwin's years aboard the *Beagle* would change his life, and transform biology.

The *Beagle* sailed from Plymouth, England, on December 27, 1831, a date that Darwin later referred to as "my real birthday" (Figure 1.4). The ship's frequent dockings enabled the young Charles to explore many new areas. His experiences were overwhelming—he hiked through a jungle in Brazil, dug up fossils in Argentina, witnessed an erupting volcano, withstood an earthquake that raised shellfish beds 3 feet above the shoreline, and studied coral reefs. Throughout the voyage, Darwin kept extensive notes; by the end of his voyage, Darwin had written 1,383 pages of notes about geology, written 368 pages of notes about zoology, taken 770 pages of notes in his diary, preserved 1,529 specimens, and labeled

3,907 skins, bones, and other specimens. He had also seen numerous interesting geological formations and new habitats (e.g., the Andes and tropical islands). His voyage lasted 58 months, 43 of which were spent in South America.

While at sea, Darwin read a book that had been given to him by Captain FitzRoy: Volume 1 of Charles Lyell's *Principles of Geology, Being an Attempt to Explain the Former Changes of the Earth's Surface, by Reference to Causes Now in Operation*. As its title suggests, Lyell (1797–1875) re-fashioned Hutton's uniformitarianism into a coherent scientific theory, documenting that ancient Earth had been molded by the same slow, di-rectionless forces that mold Earth today—namely, by earthquakes that lift and shift land, by erosion that grinds down rocks and redistributes soil, and by volcanoes that build islands. The process goes like this:

Weathering and erosion wear away the top layer of land.

This layer is washed to the ocean, where it sinks to the bottom.

These layers accumulate and form new layers of sea rock while capturing and preserving organisms trapped in the sediment.

Pressure and heat from Earth's molten core push layers of the ocean floor upward, creating new land.

Lyell claimed that "the present is the key to the past," and the fron-tispiece of his book made his point—it showed the water-marked marble pillars of the 2000-year-old Roman Temple of Serapis at Pozzuoli (near Naples, Italy; Figure 1.5). When Lyell visited the ruins in 1828, the Tem-ple's three remaining marble pillars—each some 40′ high—were still standing. The lower 12′ of each column was smooth, but the 10′ or so above this height was deeply bored by the marine bivalve, *Lithodomus* (many of the bored holes still have shells in them). The temple had been built above sea level, but the presence of the mollusks on the col-umn meant that at one time after construction the columns had been partially submerged in the ocean. The columns were then raised to their present level by the 1538 volcanic eruption that produced Monte Nuovo just northwest of Pozzuoli. The dividing line and smooth area at the bottom of each column showed how far the columns were buried in sediment, and therefore were protected from the boring mollusks (the sediments covering these parts of the columns had been excavated in 1749). Lyell's use of these columns, both as a frontispiece and as the subject of a prominent 10-page discussion near the end of *Principles*, made the Temple an icon of Lyell's uniformitarianism.

Lyell's point was obvious: evidence of the geological changes that have been shaping Earth for millennia is observable today. Following

Hutton, Lyell showed that Earth is very old, and that Earth's history has been characterized by the extinction and appearance of innumerable species (Chapter 4). This was the world that Darwin sought to explain.

Figure 1.5 The frontispiece of Charles Lyell's *Principles of Geology* shows the ancient Roman Temple of Serapis. The dark bands on the marble pillars were produced by mollusks that had drilled into them after the columns became submerged in the sea. (*Image copyright History of Science Collections, University of Oklahoma Libraries*)

In September of 1835, the *Beagle* docked for a month-long stay at a group of 13 volcanic islands called the Galápagos Islands (Figures 1.4 and 1.6). These islands, which straddle the equator, are about 600 miles west of Ecuador. The Galápagos are named for large tortoises that inhabit them (*galápago* is "turtle" in Spanish; see Figure 1.6), but Darwin did not initially find the islands to be overly interesting: "All of the plants have a wretched, weedy appearance, and I did not see one beautiful flower." The flowers he did find were "insignificant, ugly little flowers."

Although Darwin took great pains to collect insects, he also collected some songbirds called finches, which appeared to Darwin as if "one species had been taken and modified for different ends." Darwin also learned that local inhabitants could tell the home island of a tortoise just by examining its

shell. As Darwin noted later, "By far the most remarkable feature in the natural history of this archipelago [is] that the different islands to a considerable extent are inhabited by a different set of beings. My attention was first called to this fact by the Vice-Governor, Mr. Lawson, declaring that the tortoises differed from the different islands, and that he could with certainty tell from which island any one was

Figure 1.6 The Galápagos Islands (top) are volcanic islands that straddle the equator about 600 miles west of Ecuador. The view shown in the upper photo is toward San Salvador (James) Island from Bartolomé (Bartholomew) Island. The islands are home to the giant Galápagos tortoises (*Geochelone nigra*) (bottom). (*Photographs courtesy of Alton L. Biggs*)

brought." This intrigued Darwin, for it suggested that each island had its own group of organisms. Did each island really have a unique species of tortoise? Did each island have a unique species of finch? Had there been a separate creation event at each island? Darwin was initially skeptical: "I never dreamed that islands 50 or 60 miles apart, and most of them in sight of each other, formed of precisely the same rocks, placed under a quite similar climate, rising to nearly equal height, would have been differently tenanted." However, he remained puzzled. The sureties of his faith in special creation were increasingly replaced by the ceaseless questions of science.

After leaving the Galápagos, the *Beagle* sailed to New Zealand and Australia, during which time Darwin pondered the questions he had been developing about life's diversity. When Darwin got back to England in October of 1836, ornithologist John Gould (1804–1881) of the London Zoological Society examined the birds that Darwin had collected on the Galápagos Islands and told Darwin that each island *did*, in fact, house a separate species of finch.

Darwin's Life Back in England

When Darwin boarded the *Beagle* in 1831, he believed in the fixity of species and intended to spend his life in the clergy. But when he returned to England almost 5 years later, he was a changed man who wanted to discover the laws of nature. He knew that species change, but couldn't explain how they change. Back in England, Darwin set about trying to figure this out.

In September of 1838, Darwin read "for amusement" *An Essay on the Principles of Population* written in 1798 by clergyman and economist Thomas Malthus (1766–1834). Malthus argued that populations can grow exponentially (e.g., 2, 4, 8, 16, 32, 64, 128, 256, etc.; see Figure 1.7), but that resources such as food can increase only linearly (e.g., 1, 2, 3, 4, 5, 6, 7, etc.). Here's how Malthus described his idea: "Population, when unchecked, increases in a geometrical [exponential] ratio. Subsistence increases only in an arithmetic [linear] ratio. A slight acquaintance with numbers will show the immensity of the first power in comparison of the second.... As many more individuals of each species are born than can possibly survive ... it follows that any being, if it vary ever so slightly in a manner profitable to itself ... will have a better chance of survival, and thus be naturally selected." This meant that Earth's limited resources—along with war, famine, and disease—would inevitably

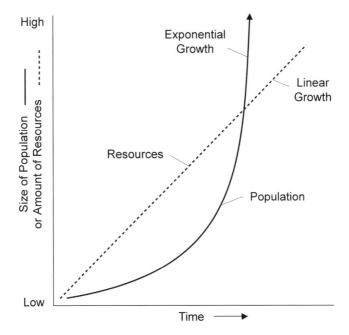

Figure 1.7 A comparison of linear and exponential growth. Thomas Malthus argued that the exponential growth of populations would ultimately outstrip the availability of life-supporting resources, thereby producing a "struggle for existence." (*Jeff Dixon*)

produce what Malthus called a "struggle for existence." (Malthus argued that famines, diseases, and wars would also control the population and "fairly be resolved into misery and vice.") Malthus claimed that starvation was designed to teach the virtues of hard work and moral behavior, and believed that the poor could do nothing but practice "moral restraint" to avoid bringing children into the world whom they could not feed. Over time, Malthus' essay became associated with repressive measures against the poor, despite the fact that Malthus advocated more humane solutions, such as using education to reduce the sizes of families. Malthus' book had a tremendous influence on Darwin: What would be the consequences of a constant struggle for existence (as Malthus proposed) that persisted for millions of years (as Lyell proposed)?

As Darwin pondered this question, he realized that Malthus' struggle throughout the history of Lyell's ancient Earth might explain the finches, the tortoises, and all the great diversity of plants and animals that he had encountered on his travels. As he noted in his autobiography, "Being well prepared to appreciate the struggle for existence ... it at once struck me that under those circumstances, favorable variations would tend to be preserved and unfavorable ones destroyed. The result of this would be the formation of a new species. Here, then, I had at last got a theory by which to work." Combining Malthus' idea with what he had seen at the Galápagos (which he called the primary source of all his views) gave Darwin a new insight. He now knew how species evolved. "Did [the] Creator make all new [species on oceanic islands], yet [with] forms like [on] neighboring continent? This fact speaks volumes. My theory explains this but no other will." Despite his confidence, however, Darwin was not yet ready to announce his discovery.

As experts continued to sort through Darwin's enormous collection of specimens from his *Beagle* voyage, Darwin began to think more and more about his radical idea. The finches that he collected from the Galápagos suggested to Darwin that new species could evolve from a common ancestor. Darwin began to think of humans not as an ultimate and special creation, but instead as merely one more species, albeit one with unusual mental powers: "It is absurd to talk of one animal being higher than another. People often talk of the wonderful event of intellectual Man appearing—the appearance of insects with other senses is more wonderful. . . . Who with the face of the earth covered with the most beautiful savannas and forests dare say that intellectuality is the only aim of the world?"

In 1837, Darwin began writing about his "dangerous" idea in a secret notebook labeled "Transmutation of Species." This notebook contains Darwin's first "tree of life," in which he depicts life not as a hierarchical ranking of "higher" and "lower" forms (as Aristotle and other naturalists had claimed), but instead as a branching tree showing shared origins (Figure 1.8). The branches of the tree did not necessarily lead anywhere; they just spread. Instead of marching up a chain or ladder as Lamarck and other naturalists had suggested, Darwin's tree showed that species evolved; in some cases, one species could give rise to many species (as had occurred on the Galápagos Islands). Although Darwin would not publish his theory for 22 years, his "tree of life" would become a metaphor for Darwin's view of how species evolve.

It is interesting to note that young Charles Darwin made the mistake that many beginning (and some experienced) scientists make—he neglected to record all the information he would later need to make full use of his data. For instance, he understood, once his finches were properly identified after his return to England, that they were all closely

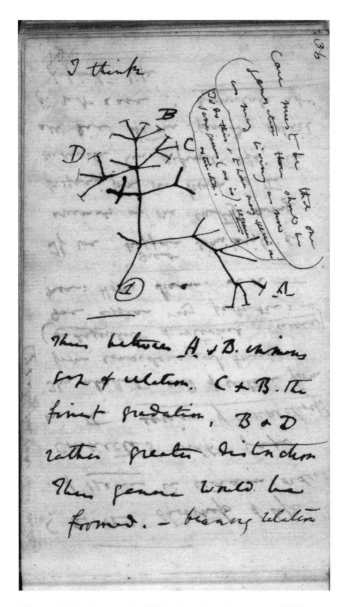

Figure 1.8 Darwin's 1837 sketch of a tree of life was drawn under the dramatic words, "I think." (*Reprinted by permission of the Syndics of Cambridge University Library*)

related. However, while at the Galápagos, Darwin had paid them little mind; because he had neglected to record the precise island from which each was taken, he could not reconstruct their probable relationships. Going back to the Galápagos was out of the question, so he asked Captain FitzRoy and other shipmates if he could borrow the Galápagos birds that they had donated to the British Museum. Darwin received six sets of bird skins, and Gould's conclusion was strengthened: Each island housed a different species of finch. Meanwhile, Thomas Bell, another Fellow from the Royal Zoological Society who had been identifying Darwin's reptiles, provided a parallel conclusion: Each island of the Galápagos chain had produced its own distinct species of iguana, a kind of lizard.

Although Darwin's theory consumed much of his time, Darwin also cherished his family. On January 29, 1839—just 5 days after he was elected Fellow of the Royal Society—Darwin married his first cousin Emma Wedgwood, and they moved into a house near Charles' brother Erasmus in London to start their family. (Charles and Emma eventually had 10 children, but only 7 would reach adulthood). Darwin was wealthy, and so was Emma, so the Darwins never had to seek employment.

While living in London, Darwin completed a book documenting his voyage aboard the *Beagle*. The book was part of a three-volume set that had a ponderous title: *Narrative of the Surveying Voyages of His Majesty's Ships* Adventure *and* Beagle, *between the Years 1826 and 1836 Describing their Examination of the Southern Shores of South America and The* Beagle's *Circumnavigation of the Globe in Three Volumes.* Darwin's book was Volume 3 of the set (another volume was written by FitzRoy). The volumes could be purchased separately, and Darwin's became a bestseller. When Henry Colburn, the publisher, reprinted Darwin's book, he gave it a grander title: *Journal of Researches into the Geology and Natural History of the Various Countries Visited by H.M.S.* Beagle *Under the Command of Captain FitzRoy, R.N. from 1832 to 1836.* Darwin's book, retitled *The Voyage of the Beagle* at its third printing, was reprinted many times. This book, one of the world's great travel books, remains a steady seller. Like Lyell, Darwin was praised as a scientist and writer ("a first-rate landscape-painter with a pen").

FitzRoy's volume, which included comments about geology and biblical history, was ignored or ridiculed; as Lyell commented about FitzRoy's last chapter, "It beats all the other nonsense I have ever read on the subject." In the years ahead, FitzRoy would manufacture a weather-station (which became known as a FitzRoy barometer) and become the first to disseminate and popularize weather lore such as "A grey sky in the morning, fine weather," and "A bright yellow sky at sunset presages wind; a pale sky, wet." However, Darwin's accomplishments would increasingly relegate FitzRoy to a historical footnote. Much to his life-long regret, FitzRoy was the man who took Darwin around the world and made Darwin's discoveries possible.

Meanwhile, the Darwins soon tired of life in "dirty" London. They moved into a large house near the village of Downe; Charles and Emma Darwin raised a family there and lived contentedly in Down House for the rest of their lives (see Down House). Despite his subsequent fame, Darwin never again left England.

DOWN HOUSE

Few historic homes intrigue biologists as much as a large house 16 miles southwest of London. There, just outside the small village of Downe, Kent, is Down House, where Charles and Emma Darwin raised their children and lived for more than 40 years. (The village was originally named Down, but later changed its name to Downe; the house kept the old spelling.) Their large house, which was built late in the 18th century, fronted a 15-acre meadow and greenhouse, where Charles did many of his experiments. Along the western edge of the property was the "Sandwalk" or "thinking path," where Darwin walked to digest his thoughts. When Darwin's friend Joseph Hooker walked with Darwin along the Sandwalk, they discussed "old friends, old books, and things distant to eye and mind."

Inside Down House, Darwin spent much time in the billiards room, where cause-and-effect relationships were much more obvious than in his theory of natural selection. Across the hall was the dining room and its bay windows that looked out on the Darwins' large backyard. A large portrait of grandfather Erasmus Darwin hangs in the dining room; lunches in this room were the focal point of the day in the Darwin home. Adjacent to the billiards room was Darwin's study, which housed his writing table, microscope, and a few biological specimens—all of which were overlooked by portraits of Josiah Wedgwood, Joseph Hooker, and Charles Lyell. Down the hall, in the drawing room, Emma played piano and challenged Charles in two games of backgammon each evening. (In a letter to Asa Gray, Darwin noted that Emma had won 2,490 games, and Charles had won 2,795.) In his study and in the large armchair by the drawing room's fireplace, Darwin sat, pen in hand, and thought out his revolutionary idea.

Charles and Emma had 10 children:

William Erasmus (1839–1914), a banker

Anne "Annie" Elizabeth (1841–1851), a daughter whose death because of tuberculosis at the age of 10 changed Darwin's views of Christianity

Mary Eleanor (1842–1842), a daughter who lived only three weeks

Henrietta Emma "Etty" (1843–1929), the editor of Emma's letters, published in 1904

George Howard (1845–1912), a mathematician and astronomer at Cambridge who became a Fellow of the Royal Society in 1879. George studied the evolution and origins of the solar system.

Elizabeth "Betty" (1847–1928), a daughter who never married and had no descendants

Francis (1848–1925), a Cambridge botany professor who published *Life and Letters of Charles Darwin* in 1887, *More Letters of Charles Darwin* in 1903, and who was knighted in 1913. Francis edited and published Darwin's *Autobiography* and was made a Fellow of the Royal Society in 1879.

Down House, where Charles Darwin lived for 40 years. Behind Down House was the "Sandwalk," where Darwin walked each day. Today, Down House is a public museum. (*Randy Moore*)

Leonard (1850–1943), a soldier and president of the Royal Geological Society, who was born well before the publication of *On the Origin of Species*, and who remained alive until after the atom had been split.

Horace (1851–1928), an engineer who built scientific instruments, and who founded the Cambridge Scientific Instrument Company. For many years the company was known as "Horace's Shop." Horace was made a Fellow in the Royal Society in 1903.

Charles Waring (1856–1858), whose death prevented Darwin from attending the public announcement of his theory at a meeting of the Linnean Society on July 1, 1858.

The Darwins loved living in Down House. As Charles noted soon after moving in, "My life goes on like clockwork, and I am fixed on the spot where I shall end it."

Soon after Charles Darwin died in 1882, Emma moved to Cambridge, after which she returned to Down House only to spend her summers. When Emma died in 1896, she was buried in the same tomb as Charles' brother Erasmus in the churchyard of nearby St. Mary's Church.

The family kept Down House until around 1900, and in 1907 it was turned into the Down House School for Girls. Down House opened as a public museum in 1929, and today a restored Down House is maintained by English Heritage. Down House is open for tours most of the year.

The "Sandwalk," where Darwin walked every day. (*Randy Moore*)

This monument is part of the wall in front of Down House. (*Randy Moore*)

In 1844, Darwin developed the ideas in his "Transmutation" notebook into a 35-page outline of "descent with modification" (as evolution was called in Darwin's day). Charles discussed his idea with a few of his close friends, most notably Joseph Hooker (1817–1911), the founder of plant biogeography, and Charles Lyell, who along with Hutton founded modern geology. After confiding in Hooker that he had discovered "the simple way which species become exquisitely adapted to various ends," Darwin likened his idea to "confessing a murder."

Darwin had witnessed the 1844 firestorm that greeted Chambers' *Vestiges of the Natural History of Creation*, and was understandably reluctant to publicly announce his theory. Instead, he continued to do research that would produce eight more books on topics ranging across insectivorous plants, earthworms, pigeons, barnacles, climbing plants, orchids, and plants' movements. Everywhere he looked, and regardless of what he studied, Darwin found evidence that supported his theory. In 1844, he expanded his 35-page outline into a 231-page (about 50,000 words) manuscript, which he stored in a hallway closet in Down House.

Darwin was often very sick from causes that remain to this day the subject of speculation and debate. Although his health was declining (e.g., he was too sick to attend his father's funeral in November of 1848), Darwin remained reluctant to announce his discovery. However, he knew that his idea was important and wanted it published, even if he should die. He wrote to his wife Emma in 1844, "I have just finished my sketch of my species theory. If, as I believe that my theory is true and if it be accepted even by one competent judge, it will be a considerable step in science . . . . I therefore write this, in case of my sudden death, as my most solemn and last request . . . that you will devote 400 pounds to its publication."

Wallace Provokes Darwin to Announce His Idea

On June 18, 1858, Darwin received a 20-page letter from young British explorer and self-trained naturalist Alfred Russel Wallace (1823–1913), who was halfway through 8 years of collecting specimens across the vast Malay Archipelago. (Wallace had written the letter in February, but it had taken 4 months to reach Darwin.) In that letter, Wallace outlined his own ideas of evolution. Darwin could hardly believe what he was reading—Wallace had come up with the same idea for evolution that Darwin had formulated (and had been secretly writing about). Wallace's moment of insight was uncannily similar to Darwin's, as shown by Wallace's description of what happened: "[Evolution] presented itself to me, and something led me to think of the positive checks described by Malthus in his *Essay on Population*, a work I had read several years before, and which made a deep and permanent impression on my mind. These checks—war, disease, famine, and the like—must, it occurred to me, act on animals as well as man. Then I thought of the enormously rapid multiplication of animals, causing these checks to be much more effective in them than in the case of man; and while pondering vaguely on this fact, there suddenly flashed upon me the idea of the survival of the fittest—that the individuals removed by these checks must be on the whole inferior to those that survived. I sketched the draft of my paper . . . and sent it by the next post to Mr. Darwin."

After consulting with friends, Darwin prepared a letter outlining his ideas, and Darwin's and Wallace's letters, along with part of Darwin's 1844 essay, were read on the evening of July 1, 1858, by Hooker and Lyell at a meeting of the Linnean Society, a leading society of professional scientists in England. Darwin did not attend the meeting (his son Charles Waring Darwin had died 2 days earlier of scarlet fever), and Wallace did not know about the meeting. The presentation generated virtually no interest among those who attended the meeting. The paper was then published under the impressive title "*On the tendency of species to form varieties; and on the perpetuation of varieties and species by natural means of selection* by Charles Darwin Esq., FRS, FLS, & FGS and Alfred Wallace Esq., communicated by Sir Charles Lyell, FRS, FLS, and J. D. Hooker Esq., MD, VPRS, FLS, & c." Nevertheless, Darwin began to work with some urgency on a book describing his idea. He finished the final chapter on March 19, 1859. As Darwin would later note, "It cost me thirteen months and ten days' hard labor." Unlike most others, Darwin understood the importance and potential impact of his idea—"It is no doubt the chief work of my life."

On November 24, 1859—22 years after Darwin had opened his secret "Transmutation of Species" notebook—John Murray Publishing of London (which had published all of Lyell's books) released Darwin's 490-page book *On the Origin of Species by Means of Natural Selection, or The Preservation of Favoured Races in the Struggle for Life.* The opening sentences of the Introduction set the stage: "When on board the H.M.S. 'Beagle', as naturalist, I was much struck with certain facts in the distribution of the inhabitants of South America, and in the geological relations of the present to the past inhabitants of that continent. These facts seemed to me to throw some light on the origin of species—that mystery of mysteries, as it has been called by one of our greatest philosophers."

All 1,250 copies of *On the Origin of Species* that had been printed sold on the first day for 15 shillings apiece. That same day, Murray asked Darwin if he wanted to make any changes before the second printing. Darwin's subsequent revisions were published in several languages that took Darwin's idea throughout the world. Darwin updated his theory and addressed critics' concerns by revising the book five times, and the final edition (the sixth edition) was published on February 19, 1872. Darwin's idea later became known as "survival of the fittest," a phrase coined in 1863 by British philosopher and economist Herbert Spencer (1820–1903). Although neither this phrase nor the word "evolution" were in *On the Origin of Species*, Darwin liked Spencer's phrase and believed that it was "more accurate" than his own explanation of natural selection.

Darwin's *On the Origin of Species* was not the first book about evolution, but it was (and still is) unquestionably the most influential. When it was published, Darwin immediately became both a celebrity and a controversial figure. Darwin stressed that his book was not a denial of God's existence, but it did challenge biblical literalism and remove humans from their pinnacle as the ultimate purpose of God's creation.

Not surprisingly, Darwin's carefully crafted book was condemned by many religious leaders, and William Whewell, Master of Trinity College at Cambridge, refused to allow it into the college library. However, many others praised Darwin's book. Darwin himself saw his idea as enlightening.

Darwin believed that "with a book as with a fine day, one likes it to end with a glorious sunset." Here is Darwin's famous conclusion of *On the Origin of Species*:

It is interesting to contemplate an entangled bank, clothed with many plants of many kinds, with birds singing on the bushes, with various insects flitting about, and with worms crawling through the damp earth, and to reflect that these elaborately constructed forms, so different from each other, and dependent upon each other in so complex a manner, have all been produced by laws acting around us.... Thus, from the war of nature, from famine and death, the most exalted object of which we are capable of conceiving, namely, the production of the higher animals, directly follows. There is a grandeur in this view of life, with its several powers, having been originally breathed into a few forms or into one; and that, whilst this planet has gone cycling on according to the fixed law of gravity, from so simple a beginning endless forms most beautiful and most wonderful have been, and are being, evolved.

Unlike many previous explanations of evolution, Darwin's theory was supported by a huge amount of evidence and included a workable, coherent, and testable mechanism that did not require a deity, miracles, or arbitrary purpose. Just as Newton had done in *Principia*, Darwin included an enormous number of detailed observations to create "one long argument" for his theory. You'll learn about the evidence for Darwin's idea in the next chapter, but here's the layout of his book. Darwin's idea, which brought historicity to biology, was based on common, everyday observations of nature.

Chapter 1 Variation Under Domestication
Darwin uses this chapter to introduce a familiar analogy from agriculture; namely, that breeders have used the extremes of nature to produce dramatic changes in domesticated animals and plants. Darwin goes to great lengths to show that we cannot distinguish between the human-caused production of new breeds and the natural appearances of new species in nature. After discussing the immense variability among pigeons, he formulates his hypothesis: "Great as are the differences between the breeds of the pigeon, I am fully convinced that all are descended from the rock-pigeon [rock dove] *Columba livia*."

✗ Chapter 2: Variation Under Nature

Darwin shows that the variation that serves as the raw material for breeders exists in wild populations (his examples include cats, cows, and dogs). In doing so, Darwin begins to submit circumstantial evidence for evolution by arguing that inherited variability is as ubiquitous in nature as in the domesticated plants and animals that he discussed in Chapter 1.

Chapter 3: Struggle for Existence

This lyrical chapter sets the stage for Darwin's most important point—natural selection, which is discussed in the following chapter. Darwin admits that "behind the face of nature bright with gladness" is a relentless and brutal struggle for existence that results from the tendency of species to overreproduce. Darwin acknowledges his debt to Malthus.

Chapter 4: Natural Selection

This chapter contains the essence of Darwin's argument. After reminding readers of the power of artificial selection, Darwin announces his new idea: natural selection. Although the phrase "natural selection" is metaphorical, there is a real process similar to it in nature—that is, nature operates on itself the way a breeder does on domestic stock. You'll learn more about natural selection in the remainder of this book, but here's a simplified version of what Darwin proposed:

- Populations can produce far more offspring than are needed to replace the parents.

- Because the resources needed to support life are limited, there is competition and struggle for existence.

- Organisms in the struggle for existence have different traits, and these different traits influence which individuals survive and reproduce. Many of these traits necessary for survival and reproduction are heritable.

- Because individuals that are best suited for survival leave the most offspring, the traits (and underlying genes) of the best-suited organisms are passed to a larger percentage of individuals in subsequent generations. The resulting accumulation of genetic change over many generations is evolution.

Darwin's argument for natural selection forms a biological parallel to Lyell's geological arguments in *Principles of Geology*; small events acting over long periods can produce large changes. This chapter includes the "Tree of Life," the only illustration in *On the Origin of Species*.

✗ Chapter 5: Laws of Variation

Darwin presents several sources of inherited variability, all of which we now know are erroneous. This was the greatest gap in Darwin's theory, and he never filled it. Darwin never knew that Gregor Mendel—an obscure Augustinian monk who enjoyed breeding plants in his monastery garden—had solved the problem and, in the process, established the foundation of modern genetics.

Chapter 6: Difficulties of the Theory

Darwin raises and dispatches several concerns about his theory (e.g., the lack of transitional forms between known species). He also presents examples of organisms that are inexplicable if one assumes that God created every species perfectly adapted to its way of life, but are expected if organisms are constantly searching for new ecological niches and if evolution takes some time to adjust a species to a new lifestyle. Darwin's willingness to expose weaknesses in his theory helped speed the acceptance of his ideas among scientists.

Chapter 7: Instinct

This chapter, which anticipates kin selection, discusses Darwin's interest in emotions; he was convinced that evolution must be able to explain animal behavior. (In 1872, Darwin developed these thoughts more fully in his book *The Expression of the Emotions in Man and Animals*.) In later editions of *On the Origin of Species,* Darwin inserts a new Chapter 7 that is entitled "Miscellaneous Objections to the Theory of Natural Selection." In this chapter, Darwin speculates that "the giraffe, by its lofty nature, much-elongated neck, fore-legs, head and tongue has its whole frame beautifully adapted for browsing on the higher branches of trees," and suggests that giraffes that could reach higher ("even an inch above the others") for food would be more likely to survive during periods of scarcity. Although Darwin bases his claim on observed facts, new observations have yielded evidence that undercuts his original idea. For example, giraffes usually eat leaves that are at shoulder height. In fact, giraffes' long necks are most useful not for browsing, but for combat. Male giraffes fighting for dominance attack each other by using their necks and armored, horned heads as clubs; winners of these contests more often mate with available females.

Chapter 8: Hybridization

Darwin addresses what many critics considered to be a flaw in his theory: the failure of distinct species, when crossed, to produce fertile offspring. Darwin claims that there is no clear distinction between varieties and species.

Chapter 9: On the Imperfection of the Geological Record

Darwin uses geology to support his theory, noting the long history of earth and that we should not expect to find all the steps of evolution preserved in fossils. Darwin attributes apparent gaps in the fossil record to incomplete sampling.

Chapter 10: On the Geological Succession of Organic Beings

Darwin presents his "succession of types" by noting that fossils undergo different rates of change. Darwin claims that new species appear, flourish, and vanish as they are replaced by other species. To Darwin, the extinct and living species are like dead and living branches of his Tree of Life.

Chapter 11: Geographic Distribution

Here, Darwin uses biogeography to bolster his previous reference to paleontology in Chapter 10. Darwin claims that if we allow for the imperfections of the fossil record, the known fossils are distributed as one would expect based on Darwin's theory. Darwin speculates that routes for dispersal were created by the rising and falling of land masses.

Chapter 12: Geographic Distribution (Continued)

This chapter magnifies the theme of Chapter 11, using island life for evidence. This chapter presents some of the evidence that Darwin gathered at the Galápagos Islands for the validity of natural selection.

Chapter 13: Mutual Affinities of Organic Beings; Morphology: Embryology: Rudimentary Organs

Darwin uses a variety of observations (e.g., rudimentary and atrophied structures as relics of once-useful organs) to develop the explanatory power of his theory. Darwin insists that assuming the existence of a divine plan adds nothing to our understanding of natural relationships. Instead, his theory alone explains why species are grouped together.

Chapter 14: Recapitulation and Conclusions

Darwin summarizes the implications of his argument, confidently adding that "young and rising naturalists" will share his vision and reject the prejudices that bind many workers to older ideas. This chapter includes Darwin's only use of the word *evolved*—it's the last word of the book.

The thesis of this book and the cornerstone of Darwin's theory is natural selection, the differential survival and reproduction of organisms. Natural selection produces adaptations, which are traits that enable organisms to survive the "struggle for existence." Darwin was convinced that just as domestic animals evolve through selective breeding (i.e., artificial selection), species in the wild evolve "by means of natural selection." Darwin explained natural selection in Chapter 4 of *On the Origin of Species*. ("I have called this principle, by which each slight variation, if useful, is preserved, by the term Natural Selection.") For Darwin, natural selection was the force that constantly adjusts the traits of future generations by sorting hereditary variations.

In *On the Origin of Species*, Darwin showed that life on Earth had indeed evolved (changed) over time, and suggested that the mechanism for evolution was natural selection. Darwin did not discuss the origin of life in *On the Origin of Species*, and referred to human evolution in one sentence that could be the understatement of the 19th century: "Light will be thrown on the origin of man and his history." Although Darwin's sentence was vague, the implications of Darwin's theory were clear to contemporary readers. They include the following concepts:

Darwin replaced the notion of a perfectly designed and benign world with one based on an unending, amoral struggle for existence.

Darwin challenged prevailing Victorian ideas about progress and perfectibility with the notion that evolution causes change and adaptation, but not necessarily progress, and never perfection.

Darwin's theory was theologically divisive, not because of what it implied about animal ancestry, but because it offered no larger purpose in nature for humanity other than the production of fertile offspring.

Darwin challenged the Providentially supervised creation of each species with the notion that all life—humans included—descended from a common ancestor. Humans are not special products of creation, but of evolution acting according to principles that act on other species.

Although Wallace always credited Darwin as the originator of the theory of evolution by natural selection, Darwin knew that his book would disturb many people; when he sent a copy of *On the Origin of Species* to Wallace late in 1859, he enclosed a note: "God knows what the public will think." However, Darwin had many defenders, most notably Harvard scientist (and evangelical Christian) Asa Gray (1810–1888) in the United States, and Thomas Huxley (1825–1895) in England. Gray, who wrote several positive reviews reconciling evolution and the Christian faith, arranged for publication of the American edition of *On the Origin of Species*. Darwin argued against special creation, but not against religion; he denied that species have separate origins, but did not deny the existence of God.

Thomas Henry Huxley is a large figure in the history of evolutionary thought. After reading *On the Origin of Species*, he proclaimed "How extremely stupid [of me] not to have thought of that!" and stepped forward to defend the idea, becoming known as "Darwin's Bulldog." As he told Darwin one week after the publication of *On the Origin of Species*, "As for your doctrines I am prepared to go to the Stake if requisite . . . I am sharpening up my claws and beak in readiness." Huxley, who coined the word "Darwinism" in 1860, claimed that Darwin's book would be considered "dangerous . . . by old ladies of both sexes," and implored his fellow scientists to take up Darwin's cause "if we are to maintain our position as the heirs of Bacon and the acquitters of Galileo." Throughout the uproar that followed the publication of his book, Darwin remained at Down

House; he was certainly interested in what was happening, but stayed out of the fray. There were no interviews, book tours, or major advertising campaigns for the most influential book in the history of biology.

After Darwin published *On the Origin of Species*, many people began to claim that humans are exempt from the forces of evolution. These claims inspired Darwin to write *The Descent of Man and Selection in Relation to Sex*, which was published in 1872. In this book, the first comprehensive theory of human evolution, Darwin unabashedly emphasized that humans, like all other species, are subject to evolution by natural selection. As Darwin noted, "We must, however, acknowledge, as it seems to me, that man with all his noble qualities ... still bears in his bodily frame the indelible stamp of his lowly origin." Darwin also stressed the importance of 'sexual selection" as a driving force for the evolution of life. If organisms are to pass their traits to future generations, they must somehow appeal to the opposite sex. Darwin's book didn't change many peoples' minds, but he was nevertheless pleased with the book.

Late in 1877, in what was one of the proudest moments of his life, Darwin received an honorary Doctorate of Law from Cambridge University. While on campus, Darwin was amused by students who dangled a monkey from a nearby building. But his health continued to decline, and Darwin suspected that his death was imminent. After joining with Huxley to convince Queen Victoria to grant the financially strapped Wallace a lifelong government pension (200 British pounds per year), Darwin made out a will in 1881 leaving money for his friends Hooker and Huxley "as a slight memorial of my lifelong affection and respect."

On April 19, 1882, Darwin told Emma to "remember what a good wife you have been" and that he "was not the least afraid to die." Later that afternoon, at age 73, Charles Darwin died in his upstairs bedroom at Down House.

The world noted the passage of Darwin and his towering intellect. For example, in Vienna, the *Allgemeine* noted that

Humanity has suffered a great loss ... Our century is Darwin's century. We can now suffer no greater loss, as we do not possess another Darwin to lose.

In Paris, the editors of *France* noted that

Darwin's work has not been merely the exposition of a system, but ... the production of an epic—the great poem of the genesis of the universe, one of the grandest that ever proceeded from a human brain.

And in nearby London, editors of the *Times* wrote that

CHARLES ROBERT DARWIN

BORN 12 FEBRUARY 1809

DIED 19 APRIL 1882

Figure 1.9 Charles Darwin is buried beneath this stone in the floor of Westminster Abbey. (*Randy Moore*)

One must seek back to Newton, or even Copernicus, to find a man whose influence on human thought . . . has been as radical as that of the naturalist who has just died. . . . Whatever development science may assume, Mr. Darwin will in all the future stand out as one of the giants in scientific thought and scientific investigation.

At a standing-room-only funeral service attended by Britain's leading politicians, clergy, and scientists (Hooker and Huxley were among the pallbearers), the choir sang a specially composed anthem from the third chapter of Proverbs to exalt Darwin's life of thought: "Happy is the man that findeth wisdom and getteth understanding. She is more precious than rubies, and all the things that thou canst desire are not to be compared unto her. . . . Her ways are ways of pleasantness and all her paths are peace." Darwin was buried in one of England's most prominent sites—London's Westminster Abbey—next to Darwin's hero, astronomer John Herschel, near Isaac Newton and Darwin's friend Sir Charles Lyell, and in the presence of kings (e.g., Henry III), queens (e.g., Elizabeth I, Mary Queen of Scots), explorers (David Livingstone), inventors (James Watt), poets (e.g., John Dryden, Robert Browning, and Alfred Tennyson), composers (e.g., George Frederic Handel), and writers (e.g., Margaret Cavendish, Charles Dickens, and Geoffrey Chaucer). As Darwin's body was lowered into the floor of the abbey (Figure 1.9), members of the choir sang "His body is buried in peace, but his name liveth evermore." They were right.

THE THEORY OF EVOLUTION AFTER DARWIN

Darwin's *On the Origin of Species* started a scientific revolution that became a turning point in biological research. A few scientists rejected Darwin's theory because of their religious beliefs and/or desires for a purposeful progression of life. For example, anatomist Richard Owen (1804–1892), a student of Cuvier who coined the term *dinosaur*, campaigned against Darwin and predicted that Darwin's idea would be forgotten within a decade. Similarly, Harvard's Louis Agassiz (1807–1873)—a protégé of Cuvier and one of the most famous scientists in America—rejected Darwin's theory, believing instead that new species arose from God's intervention after catastrophes ("We are children of God, not of monkeys"). Agassiz believed that life's history on Earth was directional ("The end and aim of this development is the appearance of man") and that the "history of the Earth proclaims its Creator." British politician and writer Benjamin Disraeli (1804–1881) summarized the views of many when he noted that "The question is this—Is man an ape or an angel? My Lord, I am on the side of the angels. I repudiate with indignation and abhorrence these new fanged [*sic*] theories."

Darwin's book also changed how scientists outside of biology approached questions. For example, when scientists before Darwin asked questions involving *why*, their answers inevitably were based on purpose (e.g., a particular structure was present because it was pleasing to a deity). In Darwin's world, purpose was replaced by function and history; a structure was there because it was (or had been) an adaptation. After Darwin, other scientists adopted similar approaches; astronomers did not seek purpose when describing the orbits of comets.

Opposition notwithstanding, by 1865 evolution had become a well-accepted scientific idea. Books, magazines, and even a few religious publications promoted evolution, and at church-affiliated Cambridge University, students were told to assume "the truth ... that the existing species of plants and animals have been derived by generation from others widely different." Darwin's idea was accepted because it made sense of isolated facts emerging from widespread disciplines such as anatomy, paleontology, and biogeography (the distribution of species on Earth). Indeed, this is the definition of a *theory*—to scientists, a theory is not some hunch, but rather an overarching and powerful concept that can explain a group of facts or observations. Darwin's idea is such a theory.

While evolution was accepted as a fact, with evidence mounting from fields as different as embryology and paleontology, the idea of natural selection as the cause of evolution entered a twilight for several decades.

Scientists came up with a variety of alternative explanations for evolution, such as a revised version of Lamarckism and an idea that species evolved according to predetermined goals. The full fruition of Darwin's magnificent idea awaited the discovery of an obscure publication about garden peas.

In Darwin's day, inheritance was thought to be the result of "blending," that is, an organism's traits were a blend of those of its parents. This model, however, was problematic for Darwin, for it meant that any trait would inevitably be diluted, and eventually disappear, from a population. In the mid-1860s, plant-breeding experiments involving garden peas saved the day when Augustinian monk Gregor Mendel (1822–1884) showed that hereditary "factors" (which we now call *genes*) do not blend, but instead are passed intact from parents to offspring. According to Mendel, each sexually reproducing organism inherits genes from its mother and father who, in turn, received genes from their mothers and fathers. These genes remain intact; they do not blend or merge with other genes. This means that the combination of genes that creates an offspring is analogous to the shuffling of a deck of cards, not the blending of ingredients to make a cake. Mendel's "particulate theory" of genetics would be important for evolutionary theory because it meant that genes (i.e., evolution's raw material, see Chapter 3) would not be diluted, but could instead remain—sometimes hidden—in a population.

Mendel published his paper in 1866 in a little known Czech journal. As a result, his ideas were overlooked for decades, and Mendel died in obscurity in 1884. However, Mendel's work was "rediscovered" in 1900, and it revolutionized genetics. In 1908, British mathematician Godfrey Hardy (1877–1947) and German scientist Wilhelm Weinberg (1862–1937) independently established what came to be known as the Hardy-Weinberg Principle, which showed (1) how genes behave in populations, (2) that populations have a vast reservoir of genetic material that can be expressed (and tested for survival) in future generations, and (3) that there is no inherent tendency for genes to disappear from a population (except by bad luck or because of natural selection). These findings were further developed in the following three decades by a group of mathematicians, population geneticists, paleontologists, and other biologists (including Thomas Morgan, George Simpson, Sergei Chetverikov, Ernst Mayr, Ronald Fisher, J. B. S. Haldane, Sewall Wright, and Theodosius Dobzhansky), all of whom found their ideas consistent with and enhanced by natural selection. This mixture of genetics, mathematics, statistics, population biology, systematics, paleontology, and Darwinian evolution came to be known as the "Modern Synthesis," and was accompanied by a wealth of new experimental data that Darwin could not

have imagined. Scientists now understood that random genetic mutations can cause changes in the traits of organisms, and that such traits—
if adaptive—can spread through a population. The power of Darwin's
idea became evident as biologists realized that all of the new discoveries
were compatible with and, in most cases, predicted by, Darwin's basic
idea (see Darwin's Bold Prediction). By the late 1950s, Darwin's theory
had become the guiding force in biological research. Darwin's theory
had transformed biology from, as one scientist put it, "a pile of sundry
facts—some of them interesting or curious, but making no meaningful
picture as a whole," to a coherent, functional idea that unified biology
and revolutionized our understanding of nature.

DARWIN'S BOLD PREDICTION

In 1862, Darwin published a book titled *On the Various Contrivances by
Which British and Foreign Orchids are Fertilized by Insects, and on the Good Effects
of Intercrossing*. In this book, Darwin makes a subtle argument for evolution
by showing how the structure of various orchids is elaborately adapted for
sexual reproduction via interactions with insects. Darwin was so confident
of this coevolution that he made a bold prediction.

The Madagascar orchid (*Angraecum sesquipedale*) has a 11″ tube, at the
bottom of which is nectar. Because the pollen of these orchids "would
not be withdrawn until some huge moth, with a wonderfully long probiscus, tried to drain the last drop" of nectar, Darwin predicted that such a
long-snouted moth must exist. That is, Darwin believed that there must be
a moth with a 11″ probiscus that would visit the plant to harvest its nectar
and, in the process, transfer its pollen to another plant. Wallace shared
Darwin's confidence: "Naturalists who visit that island should search for
the giant moth with as much confidence as astronomers searched for the
planet Neptune, and I venture to predict they will be equally successful."
Here, Wallace alluded to German astronomer Johann Galle (1812–1910),
who had searched for and found Neptune after French mathematician
Urbain LeVerrier (1811–1877) predicted its existence and position from
calculations involving the orbit of Uranus.

In the following decades, Darwin held out hope for the moth's discovery.
In 1903—41 years after Darwin's prediction and 21 years after his death,
entomologists in Madagascar discovered just such a moth: *Xanthopan morgani praedicta*. The *praedicta* was added to honor Darwin's prediction.

But the Darwinian Revolution didn't stop with the Modern Synthesis. In 1953, James Watson and Francis Crick announced the

double-helix structure of genetic material (i.e., DNA), and soon thereafter, molecular biologists began to decipher the genetic code. DNA was revealed to be a genetic "text" made of four chemical "letters," the sequence of which creates genes. The results—yet again—confirmed Darwin's theory, and allowed biologists to see at a molecular level how the DNA of various organisms had changed through time as the organisms evolved. Biologists discovered that DNA is the same in all organisms, that DNA functions the same way in all organisms, that the genetic code is the same in all organisms, that closely related organisms share more genes than do distantly related organisms, and that the products of gene expression (i.e., proteins) in all organisms are made of the same building blocks. Scientists have used this information in a variety of important ways, such as producing economically important organisms via genetic technology, diagnosing and treating diseases, producing a "molecular clock" (based on differences in organisms' DNA) to determine when species diverged, and understanding the human genome (i.e., the Human Genome Project). You'll learn about some of these applications of Darwin's theory in the concluding chapter of this book, but the point here is this: Darwin's theory had again predicted—with astonishing accuracy—the findings in a discipline (in this instance, molecular biology) that was not even imagined in Darwin's day.

Our understanding of evolution has been strengthened at every turn (Appendix 3). In the 1960s, George Williams' influential book, *Adaptation and Natural Selection*, focused attention sharply on natural selection on individuals and its role in adaptation. In the same decade, William Hamilton's explanation of how altruism could evolve opened the door to a new understanding of evolution and social behavior (see Chapter 3). Since then, more fossils have been found, more genetic material has been explored, more development has been investigated, and more behavior has been studied. At every turn, the evidence points to evolution as the source of life's diversity, and highlights the astonishing power of Darwin's idea.

SUMMARY

Throughout history, many people have proposed ideas for life's diversity. However, Charles Darwin's theory of evolution by natural selection is the only scientific theory that has consistently been supported by experimental and observational data. Darwin developed his theory over more than two decades, and today it unifies all of biology. All biologists are intellectual descendants of Charles Darwin.

2

EVIDENCE FOR EVOLUTION

In science, the strength and the value of a theory are determined by testable evidence. Does the evidence support the theory? Stated another way, does the theory explain our testable observations about nature? Does the theory enable us to make accurate predictions about life? If not, how can the theory be modified to correspond with the available evidence so that we can make more accurate predictions?

Evolution by natural selection is a powerful and breathtakingly simple idea that is based on the unavoidable conclusions that arise from common, everyday observations. You were introduced to these conclusions in Chapter 1, but we expand upon them here (Figure 2.1).

Observation: Populations can produce far more offspring than are needed to replace the parents. Indeed, if given enough resources, populations of organisms will grow exponentially (see Figure 1.7 in Chapter 1). To understand exponential growth, consider a bacterium. When this bacterium divides, the population doubles from one bacterium to two bacteria. When each of these two offspring divides, the population again doubles from two to four bacteria, and then from four to eight, from eight to sixteen, and so forth. This is exponential growth, and it can produce incredible results. To appreciate this, consider what would happen if, on the first day of the month, a wealthy relative gave you a penny and promised to double that amount every day for the rest of the month. On the second day of the month you would be given 2 cents, on the third day 4 cents, and so on. The slow start produces almost unbelievable results later; for example, on the thirty-first day of the month, you would be given more than $10,000,000.

Similar examples occur throughout life. For example, if a bacterium divides every 20 minutes and all of its offspring survive and reproduce, after 1 day there will be 4×10^{22} bacteria having a mass of

3.2 million kilograms. After 2 days, the entire Earth would be covered by a layer of bacteria (all descendants of the one original bacterium) more than 2 meters thick.

The reproductive abilities of organisms are staggering. For example,

A mature oak tree can produce more than a ton of acorns per year, an oyster can release more than 30 million eggs per year, and some tapeworms can produce more than 100,000 eggs per day. During her lifetime, a female ocean sunfish releases almost 30 million eggs.

A female housefly lays approximately 120 eggs every 2 months. If all of the offspring from one female would survive and reproduce, after 1 year of such reproduction there would be more than 5 trillion flies.

Fruit flies lay and fertilize 200 eggs every 3 weeks. If all of the descendants would live and reproduce, in 17 generations (approximately 1 year) the mass of the flies would exceed that of the Earth.

Geese live for approximately 10 years, and one pair of geese can raise a brood of eight offspring per year. If all of the descendants of this pair of geese survive and reproduce for 10 years, the family will include tens of millions of birds.

Elephants start to breed when they are near 30 years old. If one pair of elephants started to breed at age 30 and had only six offspring during the next 60 years, and if all of their descendants reproduced with a similar schedule, there would be 19 million elephants after 750 years.

Clearly, species can reproduce at unsustainably high rates.

Observation: The resources needed to support life are limited. For example, the number of plants that can grow on a particular plot of land, or the volume of freshwater in a lake, is limited. There is only so much of the food, water, and shelter needed to sustain life.

Conclusion: Because resources needed to support life are limited, the production of excessive numbers of offspring produces competition and a struggle for existence.

As mentioned in Chapter 1, Darwin's ideas about population growth and resources were strongly influenced by *Essay on the Principle of Population,* in which Thomas Malthus (1766–1834) noted in 1798 that human populations are ultimately limited by scarce resources in the environment. As Malthus noted, "necessity, that imperious all pervading law of nature, restrains [populations] within the prescribed bounds. Among plants and animals its effects are waste of seed, sickness, and premature death. Among mankind, misery and vice." (Given the not-so-cheerful nature of his conclusion, it is not surprising that Malthus' claim became known as his "Dismal Theorem.") Malthus concluded that competition eliminates weak people and leaves stronger ones to reproduce and pass their beneficial traits to off-spring. Malthus focused on humans, but Darwin extended Malthus' idea to all populations: "As more individuals are produced than can possibly survive, there must in every case be a struggle for existence.... It is the doctrine of Malthus, applied with manifold force to the whole animal and vegetable kingdoms.... There is no exception to the rule that every organic being naturally increases at so high a rate that if not destroyed, the earth would soon be covered by the progeny of a single pair."

Darwin was right. In most cases, only a fraction of offspring survive in each generation and leave offspring of their own; the others starve, are eaten, die of disease, do not mate, or are unable to reproduce for other reasons. As Darwin noted, species' reproductive rates "must be checked at some period of life."

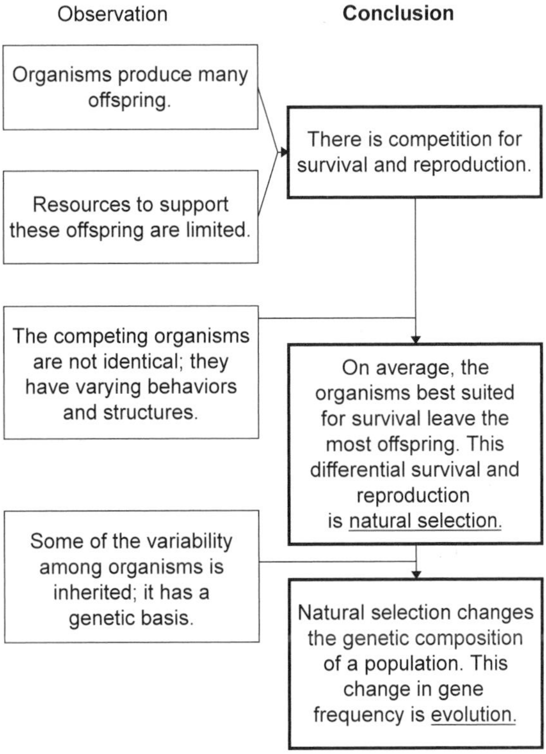

Observation	Conclusion
Organisms produce many offspring.	There is competition for survival and reproduction.
Resources to support these offspring are limited.	
The competing organisms are not identical; they have varying behaviors and structures.	On average, the organisms best suited for survival leave the most offspring. This differential survival and reproduction is <u>natural selection</u>.
Some of the variability among organisms is inherited; it has a genetic basis.	Natural selection changes the genetic composition of a population. This change in gene frequency is <u>evolution</u>.

Figure 2.1 This flowchart of evolutionary reasoning summarizes the ideas of Charles Darwin and Alfred Russel Wallace, and also includes some elements of modern genetics. (*Jeff Dixon*)

Observation: There is much variation among individuals of a population; except for identical twins, no two individuals are identical. This means that members of a population have different abilities to withstand environmental extremes, escape predators, attract mates, and so on. That is, the individuals in the competition to survive and reproduce have different traits. Some of these traits (e.g., speed, ability to withstand harsh conditions) may be advantageous, whereas other traits (e.g., poor eyesight) may be detrimental.

Conclusion: The variability among individuals helps determine which individuals survive and reproduce, and thereby which individuals leave the most offspring in the next generation. Individuals having the most beneficial traits are most likely to survive and reproduce; that is, the organisms best adapted for survival will, on average, produce the most surviving and fertile offspring. As Darwin noted, "Can we doubt that individuals having any advantage, however slight, over others, would have the best chance of surviving and of procreating their kind?" This differential rate of survival and reproduction, which Darwin referred to as "a power incessantly ready for action," is *natural selection*.

Observation: Much of the variation among individuals of a population is heritable. For example, siblings share more traits with their parents and each other than they do with unrelated members of the population. For centuries, farmers have exploited this heritability to produce offspring having desired traits.

Conclusion: Because individuals that are best suited for survival leave the most offspring, the traits (and underlying genes) of the best-suited organisms are passed to a larger percentage of individuals in subsequent generations. That is, the traits that enhance survival and reproduction are disproportionately represented in a population's succeeding generation. Over generations, the greater rate of reproduction among organisms well suited to the environment (and the lower rates of reproduction among organisms poorly suited to the environment) changes the genetic composition of the population. This change in gene frequency is *evolution*.

THE EVIDENCE FOR EVOLUTION

Evolution leaves evidence, and a mountain of evidence supports the occurrence of evolution.

The Age of the Earth

One reason that many people before Darwin rejected evolution was the claim that Earth is only about 6,000 years old. Among the earliest people to make this claim was Martin Luther (1483–1546), who launched the Reformation in 1517 when he posted the *95 Theses* on the door of the Wittenberg Cathedral. In 1541, Luther claimed that creation occurred in 3961 B.C. Virtually all subsequent chronologies followed Luther's lead and set creation near 4000 B.C.

The most influential biblical chronology was proposed in 1650 by Irish archbishop and respected scholar James Ussher (1581–1656). Like those before him, Ussher claimed that Earth was only about 6,000 years old. Ussher used ancient historical texts and biblical research to calculate that God had created Earth on Sunday, October 22, 4004 B.C., almost 3,000 years before Genesis was written. Beginning in 1701, Ussher's claim was reprinted in the margins of the King James (i.e., authorized) version of the Bible, and was accepted as fact and without question by

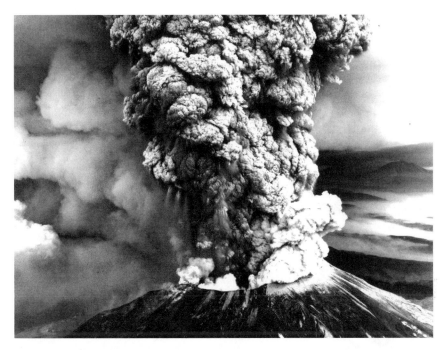

Figure 2.2 James Hutton and Charles Lyell argued that Earth's geological features had been shaped not by worldwide catastrophes, but instead by local forces that are common today, such as wind, rain, and volcanic eruptions. (*U.S. Geological Survey*)

many Christians. However, such a "young Earth" was not consistent with the process of evolution, for the evolution of life's diversity required a much longer period of time. If Earth was only 6,000 years old, evolution could not account for that diversity.

By the late 1700s, however, some scientists were questioning Ussher's claim about the age of the earth. These questions were triggered in part by attempts to satisfy the industrial revolution's huge demand for coal and metals, which prompted geologists to begin searching the earth for resources. Geology became a practical and popular profession, and geologists' searches for coal and metals led to important scientific discoveries, including previously unknown plants and animals, new ecosystems, and the fact that Earth had been changed for eons by wind, rain, running water, volcanic activity, and changes in temperature (Figure 2.2). In some areas, entire mountains seemed to have eroded away. How could this have happened in only 6,000 years? Scottish geologist James Hutton, whom you met in Chapter 1, argued that Earth was, in fact, very old, and that its geological features had been formed by the same forces that we still see at work today—namely, wind, rain, volcanic eruptions, and temperature changes.

Hutton's idea was confirmed and extended by the research of lawyer-turned-geologist Charles Lyell (1797–1875) in the early 1800s. Lyell's monumental *Principles of Geology*, which traveled around the world with Darwin aboard the *Beagle*, presented an overwhelming amount of evidence that convinced most people that Earth is, in fact, very old. Lyell's pioneering work earned him a knighthood in 1848 and laid the foundation for Darwin's ideas; as Huxley noted, "I cannot but believe that Lyell was for others, as for me, the chief agent in smoothing the road for Darwin." Lyell's work had a tremendous influence on Darwin. For example, Lyell often used the metaphor of God as the "Author of Nature," and Darwin often used this metaphor in his notes made aboard the *Beagle*. When Lyell died in 1875 at age 78, Darwin paid homage to his friend by noting that "I never forget that almost everything which I have done in science I owe to the study of his great works." Darwin modeled his *On the Origin of Species* after Lyell's *Principles of Geology*, one of the most read and talked-about books of its day. Indeed, Lyell's book established geology as the next subject to be founded fully on science (after cosmology and mechanics; chemistry and others would follow, and biology would be among the last).

Like many people who lived in Darwin's time and since, Lyell struggled with Darwin's ideas. Lyell encouraged Darwin to publish *On the Origin of Species*, but he initially rejected some of Darwin's claims. When Lyell published the first edition of *Principles of Geology* in 1830, he endorsed the idea that a special creation had produced organisms perfectly adapted to local conditions. However, Darwin's ideas helped explain how introduced species can sometimes outcompete indigenous organisms. Lyell's conclusions were consistent with Darwin's theory, and Darwin supporter Thomas Huxley correctly noted that Lyell was "doomed to help the cause he hated." Although Lyell had argued for special creation in the first nine editions of *Principles of Geology*, in 1867's tenth edition of *Principles of Geology* Lyell publicly refuted creationism and endorsed Darwin's ideas. It was no small matter for Lyell to abandon the theory that he had supported for so long in the book upon which he had established his reputation. Darwin knew this, noting that "Considering his age, his former views and position in society, I think that his action has been heroic." Lyell's *Principles* was the standard reference for geologists for almost a century, and today it remains the most important geology book ever written.

Today, scientists estimate that Earth is approximately 4.65 billion years old, and that life has inhabited Earth for at least 3.5 billion years (see

Determining the Ages of Fossils and Rocks). Such vast periods of time are extremely difficult, if not impossible, for us to understand. For example, most people can appreciate periods up to a hundred or so years; after all, we've seen photos of, and have heard grandparents' stories about, previous decades. However, we know relatively little about life during the time of Jesus (about 2,000 years ago), and even less about life 5,000 years ago when Egyptians were building the pyramids. Nevertheless, if we could go back 5,000 years to watch the Egyptians work, we would have seen only about 0.0001 percent of Earth's history (see Appendix 1).

───────────────── ౿ళ౿ ─────────────────

DETERMINING THE AGES OF FOSSILS AND ROCKS

One way scientists estimate the ages of fossils and rocks is with radiometric dating. Radiometric dating is based on the discovery in 1898 of radioactivity by Marie and Pierre Curie. Radiometric dating uses naturally occurring radioactivity as a clock, and is based on the fact that some elements have unstable isotopes, that is, the elements occur in more than one form. When these unstable isotopes become more stable by releasing radiation energy, they become a different element or a different isotope of the same element at a characteristic, unalterable rate. This is called *radioactive decay*. Such decay is not affected by environmental factors such as temperature or humidity; it proceeds at a constant rate. For example, potassium-40 decays to argon-40:

K-40 → Ar-40

It takes about 1.3 billion years for half of a sample of K-40 to decay to Ar-40. This means that the half-life of K-40 is 1.3 billion years.

Here's how radiometric dating works. When volcanic rocks form, they incorporate into themselves various elements from their surroundings. Some of these elements are radioactive, meaning that they emit radiation (i.e., energetic rays or particles) and become more stable. For example, some volcanic rocks trap K-40 as they cool. When this K-40 decays into Ar-40 in a cooled rock, the original "parent" element (K-40) is trapped in the rock, as is the "daughter" element (Ar-40). We can count both elements and, based on the ratio of K-40 to Ar-40, determine the age of the rock.

Let's examine a specific example involving carbon, an atom that is part of all living organisms. More than 98 percent of carbon atoms have an atomic mass that is 12 times that of the lightest element, hydrogen; these carbon atoms are known as C-12. About 1 percent of carbon atoms are 1 mass-unit heavier, and are called C-13. A very small percentage of carbon atoms— about one-hundredth of a percent—are heavier still, and are called C-14. Unlike C-12 and C-13, which are stable, C-14 is radioactive (i.e., unstable). Radioactive C-14 forms naturally in our atmosphere when nitrogen gas is hit by cosmic rays from space. C-14 later decays to N-14.

When animals eat food or when plants absorb carbon dioxide from the atmosphere, C-14 accumulates in their tissues. However, when these organisms die, their intake of C-14 stops because they are no longer consuming food or absorbing carbon dioxide from the atmosphere. Because no more carbon is entering the tissue, the ratio of C-14 to C-12 begins to slowly decline as the radioactive C-14 atoms decay to N-14.

The decay of C-14 to N-14 has a half-life of 5,730 years. If bones of vultures found in the Grand Canyon have only one-fourth of the expected ratio of C-14 to C-12, the C-14 in the vultures has undergone two half-lives ($1/2 \times 1/2 = 1/4$). This means that the vultures died about 11,460 years ago. Other isotopes that are used to date fossils include Uranium-235, Thorium-232, and Rubidium-87.

Scientists now have more than 40 different techniques for radioactive dating. These techniques have provided remarkably consistent results about Earth's history.

Clearly, 3.5 billion years is plenty of time for evolution to have produced life's diversity. But has it? What is the evidence?

Fossils

Knowing that Earth is so old that evolution *might* have occurred does not necessarily mean that evolution *has* occurred. Perhaps Earth and its organisms have always been like they are now, as was claimed by Aristotle and others. How could this be tested? How do scientists know that Earth and its organisms have changed over time? One excellent way of knowing this involves fossils (from the Latin word *fossilis*, meaning "dug up").

Although Darwin understood that the geologic record was one of the weakest links in his arguments (he devoted two chapters of *On the Origin of Species* to the fossil record; see Chapter 1), discoveries of fossils after the publication of Darwin's book greatly strengthened his arguments. For example, just 2 years after his book was published, biologists in southern Germany discovered the first specimen of *Archaeopteryx*, a so-called "missing link" between birds and reptiles. This organism, which lived 150 million years ago, was the earliest known bird—it had feathers and wings like a bird, as well as the jaws, teeth, a jointed lizard-like tail, and skeleton of a dinosaur.

Fossils are intriguing in themselves, and an understanding of fossils will help you appreciate the powerful nature of this type of evolutionary evidence. There are three major types of fossils:

Compression fossils form when organisms or parts of organisms are buried in water- or wind-borne sediments before decomposing. The weight of the sediment leaves an impression in the sediment, much like footprints in mud.

Premineralized fossils form when organisms or parts of organisms are buried in sediments and dissolved minerals precipitate in tissues.

Casts and molds form when organisms decay after being buried in sediment. Molds are unfilled spaces, and casts form when new materials fill the spaces and solidify.

The most common fossils are the remains of organisms preserved in sedimentary rock, the most common rocks at Earth's surface. Fossils never form in igneous rocks (i.e., rocks formed by molten rock, such as magma and lava). Humans have found fossils of more than 300,000 known species, and more are reported every week. These fossils have been found in a variety of places. For example, about 500 years ago, Leonardo da Vinci was puzzled by the fact that rocks in Italy's northern mountains contained fossilized seashells and shark teeth. How did fossilized shells of marine organisms get into rocks atop mountains far from the sea? One explanation was that the seashells were carried there by a flood, but that didn't make much sense; after all, water flows downhill (not up mountains), and besides, the shells would have been battered, if not destroyed, by a flood. A simpler and more evidence-based explanation was that the shells were once in sediments of the ocean, where they became fossilized parts of rocks. When these rocks were lifted above sea level by geological processes such as earthquakes or volcanic action, the fossils were exposed.

Few organisms become fossils. Fossils seldom form, for the soft parts of organisms are usually eaten by scavengers or microbes, and the harder parts are crushed or pulled apart. However, if conditions are favorable in those layers of sediments, the remains of the trapped organisms become fossils. The "right" conditions require that the dead animal or plant be buried rapidly in a bog or at the bottom of a sea or lake where it is protected from scavengers (this is why most fossils are of aquatic organisms). These dead organisms are covered by sediments which, after millions of years, slowly infiltrate the organisms' hard parts (e.g., teeth, bones, wood) and are converted to sedimentary rock. Geologic forces raise these rocks to Earth's surface as mountains, and rains form rivers and streams that flow down the mountains. Over long periods, these rivers and streams erode the rocks and reveal the fossils, to be discovered by Leonardo da Vinci, or even by you (Figure 2.3).

Before Darwin's time, geologists throughout the world had studied layers of rock that had been exposed by mining or erosion. What they discovered was surprising: the layers of rock were similar in different

Figure 2.3 Fossils tell the history of life on earth. This foot-long fossil is *Notogoneus osculus*, an extinct ray-finned fish that lived 38–54 million years ago in southwest Wyoming, western Colorado, and eastern Utah. When this fish was alive, the area had a subtropical climate and was home to a variety of mammals, crocodiles, and boa constrictors. (*Randy Moore*)

parts of the world. The simplest interpretation of these observations was that the layers had formed when sediments collected over long periods of time at the bottoms of ancient rivers and oceans. That is, these layers of sediment are evidence of the passing of geologic time. But what about the fossils found in those sediments?

Fossils puzzled scientists for many generations; the fossilized organisms looked like plants and animals, but the fossils were made of stone. Moreover, the fossilized organisms usually did not look like existing plants and animals. Some people believed that fossils represented God's first tries at creation, whereas other people thought that they were the animals that had not gotten onto (or that had fallen off) Noah's ark, and had therefore died in the Genesis flood. However, as scientists began to study the fossils, they made several interesting discoveries. For example (Figure 2.4),

Fossils are found in different layers of rocks that were deposited at different times in history, and the deepest (oldest) layers of rock contain the oldest fossils. This means that all of the fossils did not form at the same time.

Layers that are the same age include the same fossilized organisms. For example, fossils from the Cambrian period contain fossilized trilobites, but not dinosaurs, and fossils from the Jurassic period contain fossilized dinosaurs, but not trilobites (see Appendix 1).

Different layers of rocks can be arranged according to their ages: The oldest (i.e., lowest) layers of rock contain fossils of the most primitive organisms. In contrast, layers above them contain fossils of organisms that are similar

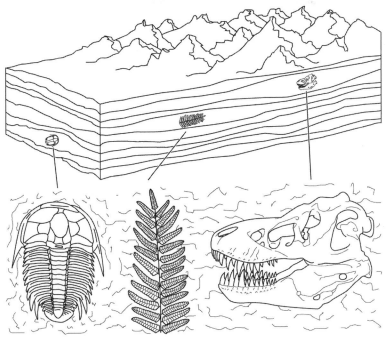

Figure 2.4 Sedimentary rock is easily visible in places such as the Grand Canyon. The deepest (i.e., oldest) layers of sediments contain the oldest fossils, whereas sediments closer to the Earth's surface contain the youngest fossils. Shown here are a trilobite, which became extinct by about 250 million years ago; a seed fern, which became extinct by about 150 million years ago; and a dinosaur, which became extinct about 65 million years ago. (top, *U.S. Geological Survey*; bottom, *Jeff Dixon*)

to these primitive forms but structurally more complex, and the youngest (i.e., uppermost) layers of rock contain fossils of organisms that resemble species alive today. Fossils of the earliest known forms of life (e.g., bacteria) appear in sediments that are about 3.6 billion years old, and younger (i.e., more recent) sediments contain a progression of multicellular organisms, fish (the first vertebrates), amphibians, then reptiles and dinosaurs, and, finally, mammals and birds. Mammals appear in the fossil record just after mammal-like reptiles having features that are intermediate between reptiles and mammals. The most recent fossils resemble organisms that live on Earth today.

Together, these observations provide strong evidence that organisms were not created at the same time, but instead have appeared on Earth over time. This chronological appearance of life makes sense if these organisms arose gradually by evolution, but would be highly improbable otherwise. As you'll see throughout this chapter, this conclusion is supported by many other types of evidence.

Some biologists tried to reconcile these new discoveries with old beliefs. For example, French anatomist and nobleman Georges Cuvier (1769–1832)—who fancied himself to be "the French Newton" and who you also met in Chapter 1—worked at a museum that had a vast collection of fossils and mummified animals provided by Napoleon's plundering armies. Cuvier studied these fossils meticulously. For example, Cuvier compared the skulls of elephants living in India and Africa with fossilized elephant skulls unearthed in Siberia, Europe, and North America. Cuvier concluded that elephants of India and Africa are separate species, and that both are different from the ancient mammoths excavated in Siberia, noting that they all differed "as much as, or more than, the dog differs from the hyena."

When Cuvier studied fossils in strata of rocks, he found that the fossilized animals in older strata were different from animals alive today. For example, Cuvier noted that marine mollusks in any one layer of rock were found only in that layer, and never in earlier or later layers. In this sense, the geological column resembled a multitiered wedding cake, with each tier having a unique flavor (i.e., a unique set of fossils). Cuvier described this finding as "the most remarkable and astonishing result that I have obtained from my research."

Cuvier's inescapable conclusion was that the history of life is recorded in layers of rocks containing fossils, that life on Earth had changed over time, and that life on ancient Earth was very different from life on Earth today. To reconcile this conclusion with his religious beliefs, Cuvier claimed that the fossilized organisms in different layers of rocks had died in a series of catastrophes, of which Noah's flood was the most

recent and dramatic; as Cuvier noted in 1796, there was "a world previous to ours, destroyed by some kind of catastrophe," and that "life on
Earth has been disturbed by terrible events." Each of these catastrophes,
Cuvier claimed, eliminated some species that dated from creation, and
after each catastrophe, organisms repopulated the world. According to
Cuvier, the species that repopulated the world were not new species;
instead, they were existing species for which scientists had not yet found
fossils, and these species remained unchanged until they, too, became
extinct as a result of the next catastrophe. Cuvier's explanation enabled
him to accept geologists' discoveries without challenging a literal interpretation of Genesis. Ironically, Cuvier refused to accept the insights
provided about life's history by his own work. Although Cuvier could
offer no explanation of the catastrophes he proposed, he vehemently
opposed evolution until his death in 1832.

Cuvier was a very influential scientist, but it wasn't long before some
people began to question his reasoning. For example, Cuvier believed
that each layer of sediment described a catastrophe, but explorers soon
discovered that there are *hundreds* of distinct layers of sediments and
fossils. Moreover, some organisms appeared in only one layer of sediment, while others appeared in several layers, and human bones were
found in layers much older than 6,000 years. These observations were
not consistent with Cuvier's interpretation.

In summary, when many of us think of fossils, we think of a huge,
threatening, lizard-like monster, with a head full of teeth, greeting us as
we walk into a museum. However, most fossils are much smaller and less
daunting. In fact, most organisms never become fossils at all. As a result,
the fossil record remains incomplete. Most fossilized organisms are from
marine habitats, or from low-lying, wet, terrestrial ones, and do not retain
any tissues that might allow analysis. Despite these limitations, however,
the fossil record is very diverse and enables us to piece together the basic
history of life on Earth. We know that fossils are the preserved remains
and impressions of organisms that lived in the past; they are the pages of
life's history book that tell us that life changes and has not always been
like it is now. Thanks to the work of many scientists following the lead of
Hutton and Lyell, we know that most geological changes are slow and
gradual, and not catastrophic, and that contrary to the claims of Cuvier
and others, there is no scientific evidence for any worldwide floods. The
fossil record is strong evidence for the occurrence of evolution.

Extinctions

Before Darwin, many people had noted that living organisms are well
adapted for the lives that they lead. For example, biologists observed

that plants that live in deserts often have succulent, water-storing roots and stems—these organisms live where water is scarce, and their ability to store water is important for their survival. Many people believed that these correlations of structure and function were evidence that the organisms were created perfectly by a deity for life in their particular environments. However, biologists soon discovered evidence that caused them to question this assumption. Atop the list of these inconsistencies were extinctions. Indeed, *Tyrannosaurus* no longer stalks its prey across North America, trilobites no longer crawl on ocean floors, and pterosaurs no longer fly overhead. If organisms had been created perfectly, why had more than 99 percent of all species that had lived on Earth gone extinct?

Although the rates of extinction have varied, there have been several periods during which extinctions were very common. These peak times of extinctions are called *mass extinctions.* For example, the mass extinction that occurred near the end of the Cretaceous (about 65 million years ago) eliminated 60–75 percent of the species alive at that time. This event affected virtually every group of plants and animals. Some groups—most notably the dinosaurs—were driven to extinction, but a few (e.g., crocodiles) were not affected very much.

What caused these extinctions? You'll learn more about mass extinctions in Chapter 4, but for now, it is important to recognize that mass extinctions seem to be caused by dramatic environmental change. Although mass extinctions occur when the majority of species go extinct, there is also a continuing "baseline" extinction rate; that is, we suspect that extinction of individual species occurs more or less continuously at a low level (Figure 2.5). Today, increasing numbers of species are threatened with extinction, mostly because of human activities, which are altering habitats at unprecedented rates.

Extinctions represent biological failures; these organisms have disappeared from the Earth. They are also evidence against the "perfect" creation of life. In a perfectly created world, extinctions defy explanation; after all, why would an all-powerful deity create such failures? However, in Darwin's world, extinctions are to be expected. Extinct organisms are merely those that could not adapt to competitors, changing climate, or other forms of environmental change.

Artificial Selection

For centuries, humans have used artificial selection—that is, selective breeding—to produce domesticated plants and animals having traits that we want. For example, we've used selective breeding of the same

Figure 2.5 Virtually all species that have lived on Earth are now extinct. Among the most famous of these extinct organisms are dodo birds, dinosaurs, and American passenger pigeons (shown here). The last known American passenger pigeon died at the Cincinnati Zoo in 1914. (*U.S. Fish and Wildlife Service*)

species of cow (*Bos taurus*) to produce breeds of cattle that yield large amounts of steak and roast (e.g., Hereford and Angus cattle) and other breeds that produce milk (e.g., Jersey cattle). Similarly, we've used selective breeding to produce cabbage, broccoli, cauliflower, Brussels sprouts, kale, and kohlrabi from the same species of wild mustard (*Brassica oleracea*; Figure 2.6), and the selective breeding of other species to produce hens that lay more eggs, crops that can withstand drought, and countless varieties of pets and other animals—for example, more than 200 breeds of horses, 800 breeds of cattle, and 700 breeds of dogs. A Saint Bernard and a Chihuahua have remarkably different features, but they are both the same species (*Canis familiaris*) and share the same gene pool.

In medicine, we've used artificial selection to produce attenuated (i.e., weakened) viruses for use as vaccines against diseases such as tuberculosis, typhoid, measles, poliomyelitis, and smallpox. (These viruses are strong enough to elicit the production of antibodies and lasting immunity, but not strong enough to elicit disease.) We've even crossed some species to produce hybrids that are new species (e.g., Kew primrose).

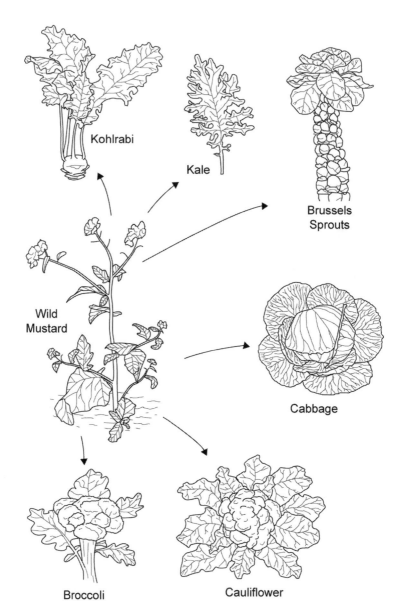

Kohlrabi

Kale

Brussels
Sprouts

Wild
Mustard

Cabbage

Broccoli

Cauliflower

Figure 2.6 Artificial selection is a human-directed and accelerated version of natural selection, the mechanism for evolution
proposed by Charles Darwin. Artificial selection can produce a variety of economically important plants and animals. Shown here
are the different results of artificial selection with wild mustard
(*Brassica oleracea*). (*Jeff Dixon*)

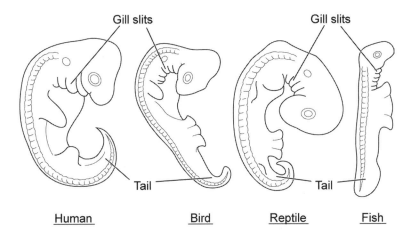

Gill slits   Gill slits

Tail   Tail

Human   Bird   Reptile   Fish

Figure 2.7 Embryos show evolutionary relationships. The embryos
of these animals share many features (including a tail and gill slits),
suggesting that these animals share a common ancestor. (*Jeff Dixon*)

Artificial selection is based on the inherent variability that occurs in natural populations. This naturally occurring variation among
organisms was an important part of Darwin's theory (Chapter 1); he
argued that nature, like farmers and dog breeders, selects organisms for
reproduction, and that the basis of this selection is the natural selection
of organisms' adaptations for survival and reproduction. Wild species
can diverge from their common ancestors just as domestic species do.

Embryology

The embryos of related organisms develop similarly. To appreciate
this, consider the embryos of the vertebrates shown in Figure 2.7. The
embryos of all these organisms have tails; these tails become swimming
structures in fish, but do not develop in humans. Similarly, the limbs
of all tetrapods (i.e., animals having four limbs such as arms and legs;
tetra = four, *poda* = foot) develop from limb buds in similar ways, and embryos of all vertebrates also have gill-like branchial arches. In fish, these
arches become gills, but in primates these structures do not become
gills; instead, they become modified for other functions, such as the
Eustachian tubes that connect the middle ear with the throat in
humans. We will see how these similarities are used in understanding
evolutionary relationships in Chapter 4. For now, note that these
embryological and developmental similarities are difficult to explain if
one assumes that each was created independently. If all vertebrates were
created independently, why do human embryos have tails and branchial

arches like our ancestors? A simpler explanation for the similar developments of the embryos of all vertebrates is that they share a common ancestor.

Evolution is conservative; instead of starting from scratch, it builds on what has come before. For example, instead of producing new developmental genes, it is more likely that natural selection will favor modifications of preexisting genes. This is why embryology provides such strong evidence for evolution. As Darwin noted, an organism's embryo is "by far the strongest single class of facts in favor of change of forms."

Biogeography

In the previous chapter, you learned how Darwin's ideas about life's biodiversity were influenced by his observations while aboard the *Beagle*, especially those of finches on the Galápagos Islands. Darwin wondered why the Galápagos finches resembled South American mainland finches more than those on islands that had similar habitats but that were much further away. To Darwin, the simplest explanation was that the Galápagos species evolved from their South American ancestors. This conclusion was based on *biogeography*, which is the geographic distribution of species.

The study of biogeography began centuries ago when Europeans began to explore the world—their journeys to Africa, the New World, Asia, and Pacific islands are important parts of today's history books. In addition to discovering that the world is much bigger than Europe, the explorers also discovered thousands of plants and animals that they had never seen before. Attempts to understand the distribution of these new organisms prompted many questions. For example, in 1832 Darwin collected fossils of glyptodonts, which were giant armadillo-like animals that lived in Argentina, but are now extinct. Darwin wrote to his mentor John Stevens Henslow that "Immediately when I saw them I thought they must belong to an enormous Armadillo, living species of which genus are so abundant here." Today's armadillos are the only animals that resemble glyptodonts, and armadillos live in environments similar to those in which glyptodonts lived. If armadillos and glyptodonts were created at the same time, why are only armadillos still alive? One explanation is that they were not created at the same time and that glyptodonts are ancestors of armadillos.

Explorers also noted that lands with similar climates (such as Australia, South Africa, and Chile) had many unrelated plants and animals, indicating that biological diversity is not controlled entirely by climate and environment. Indeed, diverse environments in a region are not typically filled with the same species that occupy similar environments in other parts of the world, but by different species that have evolved independently in those areas. These distributions of organisms are to be expected if they arose from common ancestors that lived in those areas.

There were many other questions facing the explorers. For example, why are some plants and animals found only on particular islands and nowhere else? Why do similar looking plants and animals often live in similar kinds of environments, even when these similar environments are on different continents? If all species were created in one place at one time, how did each get to its new location? And why aren't all of the plants and animals still living where they were originally created? How did each plant or animal become so well adapted to its current environment? Cuvier and many other early scientists struggled with this, for their observations were not consistent with a literal reading of the static creation story of the Bible.

Today we know that related species have similar body plans. For example, when the *Beagle* visited the Galápagos Islands (see Figure 1.6 in Chapter 1), the plants and animals there resembled, but were not identical to, those on the nearest mainland (in this instance, the western coast of South America). These similarities of organisms on the Galápagos Islands with those of nearby South America were especially striking because the Galápagos are dry and rocky, whereas the nearest part of South America is humid and has a lush tropical jungle.

Whereas Cuvier was troubled by biogeography, Darwin's theory of evolution by natural selection worked well to explain these observations. Darwin concluded that instead of organisms having been specially created on each island, species from neighboring continents migrated or were carried to the islands, where natural selection led to the evolution of adaptations to the new environments. Today, for example, we know that the Galápagos finches descended from a group of birds called tanagers that now live in South and Central America and the Caribbean Islands. Darwin's explanation for the distribution of these organisms was simple and convincing: organisms are found in particular environments because they evolved from ancestors that lived in those environments. That is, the geographic distribution of Earth's organisms fits the predictions of evolution.

Comparative Anatomy

Evolution is also supported by comparative anatomy, which is the comparison of structures across species. For example, let's return to the tetrapods. These animals have pentadactyl (i.e., 5-digit) limbs. Although these vertebrates live in many different kinds of environments and use their limbs for many different tasks, there is no clear functional or ecological reason why all of the vertebrates should have 5-digit limbs. Why not 3-digit, 14-digit, or 82-digit limbs? The structural similarities common in nature are hard to explain if the organisms originated independently of each other; after all, an independent origin would not require pentadactyl limbs or, for that matter, any other shared traits. However, the presence of 5-digit limbs in all of these organisms, despite the fact that such structure is not functionally necessary, makes sense in the light of evolution because it is evidence of their common ancestry—all tetrapods descended from a common ancestor that had pentadactyl limbs.

These shared structures are *homologous*, meaning that although the structure and function of the bones have diverged, they are derived from the same body part present in a common ancestor. To appreciate this, consider the forelimbs of mammals, such as bats, dolphins, seals, whales, and humans (Figure 2.8). All of these and other vertebrates have similar forelimbs. Richard Owen, a prominent English anatomist and foe of Darwin, believed that these homologies revealed the handiwork of a creator who saved time and work by varying a basic theme. Darwin, however, thought that Owen's explanation was useless, noting that "nothing can be more hopeless than to attempt to explain this similarity ... by utility or by the doctrine of final causes" and that the similarities are "inexplicable" by traditional views of creationism. After all, if these organisms had originated independently of (and were therefore unrelated to) each other, then one would expect their forelimbs to have different structures. However, all vertebrates' forelimbs—a bat's wing, a squirrel's leg, a dolphin's flipper, and a human's arm—have similar structures despite their different functions, sizes, and shapes. Although there is no overriding mechanical reason for these limbs to be like one another structurally, they always develop in the same way, are always in the same position on the animals' bodies, and are always made of the same arrangements of the same tissues. Similar arrangements of parts of forelimbs are also present in ancestral reptiles and amphibians, and even in the first fishes that came out of the water onto land hundreds of millions of years ago. If each of the structures was created specially for each species, why are these structures so alike in such different organisms? Why would the flipper of a dolphin have the same basic arrangement of bones as the leg of a dog or the wing of a bat? As Darwin noted, "What can be more curious than that the hand

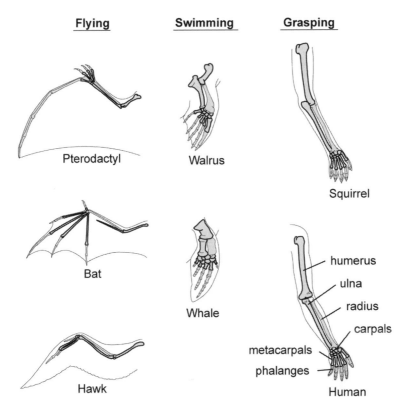

Flying **Swimming** **Grasping**

Pterodactyl

Walrus

Squirrel

Bat

Whale

humerus

ulna

radius

carpals

metacarpals

phalanges

Hawk

Human

Figure 2.8 Homologous structures. Although these limbs have different functions, they all contain the same set of bones, inherited through evolution from a common ancestral vertebrate. The bones are shown in different shades to highlight their similar arrangements in the different animals. (*Jeff Dixon*)

of a man, formed for grasping, that of a mole for digging, the leg of a horse, the paddle of a porpoise, and the wing of a bat, should all be constructed on the same pattern, and should include the same bones, in the same relative positions." In fact, not much "new" has happened to these limbs—the large differences in their structure and function result from a simple regulatory shift of a basic shared pattern; in different animals, different parts grow at different rates.

Structural similarities also occur in plants. Although most plants use leaves for photosynthesis, many plants have leaves that have been modified by evolution for functions other than photosynthesis. For example, a cactus spine and a pea tendril have different appearances and different functions; a spine protects the cactus stem, and a tendril helps support

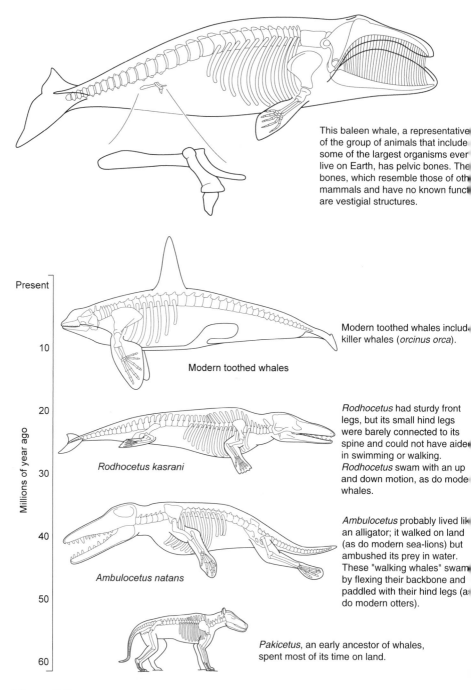

This baleen whale, a representative of the group of animals that include some of the largest organisms ever live on Earth, has pelvic bones. The bones, which resemble those of oth mammals and have no known funct are vestigial structures.

Modern toothed whales includ killer whales (*orcinus orca*).

Modern toothed whales

Rodhocetus had sturdy front legs, but its small hind legs were barely connected to its spine and could not have aide in swimming or walking. *Rodhocetus* swam with an up and down motion, as do mode whales.

Rodhocetus kasrani

Ambulocetus probably lived lik an alligator; it walked on land (as do modern sea-lions) but ambushed its prey in water. These "walking whales" swam by flexing their backbone and paddled with their hind legs (a do modern otters).

Ambulocetus natans

Pakicetus, an early ancestor of whales, spent most of its time on land.

Present

10

20

Millions of year ago

30

40

50

60

Figure 2.9

the climbing stem of the pea plant. However, both are modified leaves; they develop in the same way as leaves and consist of the same parts as leaves. Such modifications of organs used for different functions are an expected outcome of a common evolutionary origin.

However, some structural similarities do not result from common ancestry. Instead, these structures result from *convergent evolution*, in which natural selection causes nonhomologous structures that have similar functions to resemble one another despite their dissimilar origins. The development of wings on birds and insects is an example of convergence. Although birds and insects have wings, this similarity arose not from the evolutionary modification of a structure that birds and insects inherited from a common ancestor, but rather from the evolutionarily independent modification of two different, nonhomologous structures that eventually gave rise to superficially similar structures (i.e., wings). That is, natural selection favored flight in birds and insects, and the two groups independently evolved superficially similar structures that are useful for flight. These superficially similar but nonhomologous structures are *analogous structures*. Analogous structures usually have a different internal anatomy because they are not derived from structures of a common ancestor.

Structures sometimes have no use at all. These so-called vestigial structures provide even more support for evolution.

Vestigial Structures

The widely held view in Victorian England during Darwin's time that organisms were the perfect creation of a deity was also undermined by discoveries that organisms have useless parts that at one time may have been useful in an ancestor. Such useless structures are common in nature:

Some snakes (for example, boa constrictors) have rudiments of a pelvis and
    tiny legs buried in their sleek bodies, left over from an ancestor that had

←——————————————————————————————————

Figure 2.9 Although Ishmael of Herman Melville's *Moby Dick* claimed that he took "the old-fashioned ground that the whale is a fish, and call upon holy Jonah to back me," he was wrong. Whales (and porpoises and dolphins) are not fish; they are mammals that descended from hippopotamus-like mammals that lived on land. These ancestors had legs attached to their shoulders and hips, which in turn were attached to the spine. Today, whales' tiny pelvic bones are evidence of their land-based ancestors. Such vestigial structures are common in modern organisms. (*Jeff Dixon*)

legs. In 2006, biologists in Argentina discovered fossils of two-legged snakes that lived 90 million years ago.

Many species of flightless beetles have wings that cannot function because of wing-covers that are fused together and never open, a reminder that flightlessness is an acquired feature in beetles.

Pigs have vestigial toes that do not touch the ground.

Wingless birds such as kiwis have vestigial wing bones.

Many blind, cave-dwelling, and burrowing animals have nonfunctional, rudimentary eyes.

The single-celled parasite that causes malaria carries within itself the remnants of a chloroplast.

Modern whales have no hindlimbs or bony supports, but they do have small pelvic bones that have no apparent function and are not anchored to any other bones. Why? Because whales' ancestors lived on land. We know this because scientists have discovered fossils in the hills of Pakistan that show—step by step—how hippopotamus-like terrestrial organisms took to the sea and evolved into the first whales (Figure 2.9). Whales' pelvic bones are the evolutionary remnants of their ancestors' lives on land.

Clearly, organisms are living museums in some ways, full of useless but intriguing parts that are remnants of and lessons about the organisms' evolutionary histories. Humans are no different in this regard. As evolution predicts, humans also have many useless parts; more than 100, in fact. For example,

Why do humans have an appendix, which has no known function? The great apes (our closest relatives) have a larger version attached to their guts; it houses bacteria used in digesting the cellulose eaten by the primates. The appendix in humans may have once been helpful in processing different kinds of food, but today it has no known function. Indeed, more than 300,000 people per year in the United States have their appendix removed, with no long-term effects.

Why do humans have body hair? Tiny muscles at the base of hairs helped our ancestors move their hair and thereby regulate their insulation. However, in humans these muscles have no known function; all they do is give us "goosebumps."

Humans' ears have three extrinsic muscles that enable us to move our ears independently of our heads. In animals such as dogs and rabbits, orienting ears could help the organisms detect predators. In humans, however, these muscles do little more than enable us to wiggle our ears.

Humans don't have tails, but we do have bones similar to those that form the tails of other animals. These bones, which are the terminal 3–5 vertebrae, are fused into a tail-like structure (3–12 cm long) called the coccyx and are what is left of a tail that other mammals still use for balance and communication. We also have remnants of muscles for moving our vestigial tails.

At the molecular level, our genome contains numerous sequences of nonfunctional DNA, including pseudogenes, which are silent, nontranscribed sequences that are presumably similar to the functional genes in our ancestors.

From a special creation point of view, vestigial structures are difficult to explain. Why would a deity create an organism that has useless parts? However, vestigial structures are consistent with Darwin's suggestion that they were useful in the species' ancestors. The more sensible explanation for vestigial structures is not that they were specially created, but that they are the remnants of structures that were present and functional in ancestral organisms. They are the "evolutionary baggage" carried by virtually all species.

Molecular Biology

In 1949, zoologist A. S. Romer noted in *The Vertebrate Body* that structural homologies such as those shown in Figure 2.8 "might well depend upon the degree of identity of the genes concerned in their production, but . . . it is improbable that our range of knowledge will ever be broadened to the necessary degree." Were he still alive, Romer would be surprised to learn that the most impressive data gathered during the past 50 years supporting evolution have come from molecular biology. Indeed, all organisms—ranging from bacteria and fungi to plants and animals—use the same genetic code and the same molecular machinery. For example,

1. DNA stores and transmits genetic information the same way in all organisms. *Genes* are portions of DNA that code for specific traits. You will learn a bit about DNA and genes in the next chapter. For now, it is important to understand that animals share much of the hereditary material that we call genes. Indeed, the same set of 25,000–30,000 genes is present in most animals.

2. The genetic code is the universal homology; it is the same in all organisms. DNA in all organisms is made of the same chemicals and is converted to proteins the same way in all organisms. No other code has been found. Similarly, all organisms are made of cells; no other fundamental structure has been found.

3. Proteins in all organisms are made of roughly the same 20 building blocks. These building blocks are called *amino acids*.

4. The molecule ATP (an abbreviation for "adenosine triphosphate") is the universal form of immediately usable chemical energy. All organisms use ATP to do work.

5. The proteins that perform similar functions in different species have similar structures, despite their presence in different types of organisms, and more closely related organisms share more such similarities. For example, a variety of proteins in chimpanzees and humans have identical structures. Proteins in more distantly related organisms such as pigeons and fruit flies are similar, but not identical.

6. Different species may have different genomes, but the differences between the species often result from differences in the behaviors of genes that regulate the expression (or activation) of other genes. That is, the differences in species such as chimps and humans (whose genomes are more than 98 percent alike) may not be entirely due to unique genes, but instead to how and when the genes turn on and off. The underlying similarities in these species are most evident in the embryo and juvenile stages (i.e., before the differential expression of genes has occurred). This explains, for example, why the human embryo has a tail and gill-like slits (Figure 2.7), as well as why the juvenile form of a frog is a fish-like tadpole.

Advances in molecular biology during the past few decades have also produced a so-called *molecular clock*, which is based on the fact that mutations—the primary way that variation is introduced into a gene pool—may accumulate in a given piece of DNA at a reliable rate. For example, the gene that tells the body how to make the protein cytochrome *c* (which helps most organisms extract energy from their food during cellular respiration) changes at a relatively constant rate (Figure 2.10). Comparing the structures of this gene in different organisms gives us a "molecular clock" that we can use to estimate how long it's been since the species shared the same ancestor.

Molecular clocks are useful for estimating the approximate dates that species have diverged from their ancestors. For example, in 2006, biologists used this clock to trace the American domestic cat and its cousins—including lions, leopards, and panthers—to a common ancestor that lived in Central Asia more than 10 million years ago. The "roaring cats" (e.g., lions and panthers) first evolved from the common ancestor, and ocelots, lynxes, and pumas followed. The ancestors of domestic cats developed about 6.2 million years ago. The cat family has survived at least

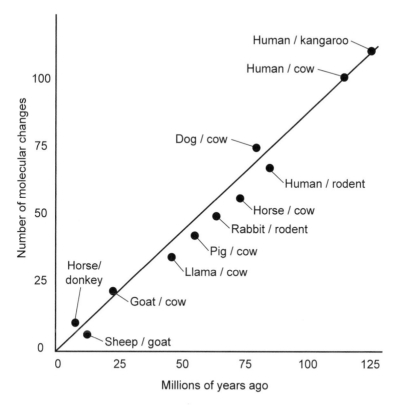

Figure 2.10 Genes function as molecular clocks. When the time since each pair of these organisms diverged is plotted against the number of molecular differences in their genes for cytochrome *c*, the result is a straight line. This suggests that the gene for cytochrome *c* is evolving at a constant rate, and therefore that molecular changes in this gene can be used to estimate how long ago species diverged. (*Jeff Dixon*)

10 intercontinental migrations and major geological events, and today cats inhabit all continents except Antarctica.

All of these observations make sense in the light of evolution, which tells us that organisms share a common ancestor. However, there's more. For example, humans have 23 pairs of chromosomes, which are large chunks of DNA (see Chapter 3); 18 of these pairs are virtually identical to their counterparts in the great apes. Similarly, human DNA is 98.3 percent identical to that of chimpanzees, 98.2 percent identical to that of gorillas, and 96.7 percent identical to that of orangutans; it is 93.0 percent identical to the DNA of rhesus monkeys (which are not apes), and 85 percent identical to the DNA of mice. Human proteins, when compared with chimpanzee proteins, usually differ by one or two amino acids, if at all.

Taken together, these observations strongly support the hypothesis of a shared common ancestor, and are hard to reconcile with a special creation. If species were each created specially and uniquely, why do they all share these fundamental similarities?

WE CAN WATCH EVOLUTION HAPPEN

Charles Darwin never tried to watch evolution happen because he believed that it occurs too slowly for us to see it during our lifetimes. As he wrote in *On the Origin of Species*, "We see nothing of these slow changes in progress, until the hand of time has marked the lapse of ages." Most scientists after Darwin also assumed that they, like Darwin, could not watch evolution in action.

This changed in the 1970s, when Peter and Rosemary Grant returned to the Galápagos Islands to study Darwin's finches. Unlike Darwin, who spent only a few weeks in the Galápagos in 1835, the Grants have spent several months there every year for more than 30 years. During this time, the Grants have done what Darwin couldn't: They have watched natural selection produce evolution, right before their eyes.

When they returned to the Galápagos Islands, the Grants assumed that they, like Darwin, would have to infer the evolution of the birds based on the birds' distribution and traits throughout the archipelago. Their work centered on Daphne Major, an island that is 5 miles from its nearest neighbor, and where environmental conditions often fluctuate wildly from year to year. Here, the Grants have seen what many believed could never be witnessed: The finches' beaks adapt to environmental changes from generation to generation.

Populations of finches on Daphne Major have a range of body sizes and beak shapes. Finches having large beaks can break open and eat large, hard seeds; finches having small beaks cannot, but can eat smaller, softer seeds. During dry years, all seeds are in short supply (thereby accelerating the competition among finches to find food), and harder seeds are more abundant than softer seeds. In dry years such as 1977, the population of medium-size ground finches (*Geospiza fortis*) plunged from 1,300 to 300 birds, and birds with larger, stronger beaks were most likely to survive; that is, the average size of beaks on the island increased as a result of the mass starvation of the smaller-beaked finches. More specifically, the finches that ultimately survived the drought had beaks that were 6 percent larger than nonsurvivors before the drought, and were thereby able to eat the large, tough seeds that were most abundant during the drought. As would be predicted from natural selection, the survivors' offspring had beaks 4–5 percent larger than nonsurvivors. That is, the drought preferentially selected genes from the starting population that brought about larger beaks, resulting in the population of finches evolving larger beaks. In wet years such as 1984–1985, there was a disproportionate abundance of smaller, softer seeds, and finches with smaller beaks were more likely to survive and reproduce. Clearly, there was no evolution toward some generally superior finch; on the contrary, the beaks of finches evolved merely in response to environmental conditions.

Although the Grants' work with Darwin's finches is the most convincing study of evolution in action ever conducted, there are many other examples of evolution that can be watched:

In Alberta, Canada, trophy hunters have killed the biggest bighorn rams for almost four decades. This preferential killing of the largest rams means that it is advantageous to be a small ram, which has a better chance of surviving and passing its genes to offspring. As would be predicted, the frequency of small adult males with small horns has increased during the past decades.

On the island of Trinidad, guppies live in streams with their predators (pike cichlids). Although female guppies choose larger, brightly colored males as mates, these males are easily spotted by their predators. In upstream water that is too shallow for the predators, male guppies are large and brightly colored. However, in the deeper downstream waters where predators live, most males are smaller and dull-colored. If predators are removed from the deeper waters—that is, if selection against being large and colorful is removed—the population sports larger and more brightly colored males within just a few generations, because those are the males that will leave more offspring in the absence of predators. If predators are then re-introduced, the frequency of smaller and duller males increases within a few generations. In each instance, these changes reflect the effect of natural selection, which favors fish with higher reproductive success, just as Darwin's theory would predict.

The moth *Heliothis virescens* attacks cotton plants in the southeastern United States. In May 1987, a new pesticide was introduced that killed 94 percent of the moths. By the end of the year (i.e., a few generations later), the same poison killed only 39 percent of the moths. The moths evolved resistance to the pesticide (see Chapter 3).

There are numerous other examples. The shell shapes of the marine snail *Littorina obtusata* have thickened in the past century, presumably in response to heavy predation by crabs. The beaks of soapberry bugs (*Jadera haematoloma*) have gotten longer in response to the invasion of a non-native plant species having large fruit (see Chapter 4). The bills of scarlet honeycreepers (*Vestiaria coccinea*) have gotten shorter and the birds have switched to other sources of nectar as their favorite flowering plants began to disappear. The list of examples goes on and on.

EVOLUTION, LIFE, AND DEATH: THE EVOLUTION OF DRUG RESISTANCE

Populations of bacteria such as *Staphylococcus* exhibit a great deal of variability. In some organisms, this variability involves differences in particular molecules in cell walls, which are the structures that surround and protect bacterial cells. These molecular differences can make the bacterium resistant or susceptible to a particular antibiotic. In the presence of an antibiotic such as penicillin, the tiny percentage of bacteria whose natural variability makes them resistant to the antibiotic are not affected by the antibiotic. However, other individuals in the population are killed by the antibiotic. This differential survival of organisms in a population means that the only organisms that remain after exposure to penicillin are those that are immune to the antibiotic. These organisms continue to reproduce, and most of these offspring—like the resistant parental bacterium—will be resistant to the antibiotic. The overall result is that the population changes over time. The original population had only a few individuals that were resistant to the antibiotic, but the selection pressures changed when the antibiotic was introduced into the environment. As a result, only the resistant bacteria could reproduce, and their antibiotic-resistant offspring formed the new generation. This is evolution.

The evolution of antibiotic resistance occurs relatively fast because bacteria have short regeneration times (some species can divide every 15 minutes or so). The evolution of drug resistance occurs especially fast in hospitals because they house sick patients whose conditions make them more vulnerable to infections, and who therefore require more antibiotics. The resulting heavy use of antibiotics in hospitals hastens the evolution of antibiotic resistance.

To appreciate this, consider that penicillin, the first antibiotic to be discovered, was hailed as a "miracle drug" because it was remarkably effective at killing bacterial pathogens. When penicillin was introduced in 1943, many people believed that infectious diseases would soon be a thing of the past. Indeed, the U.S. Surgeon General later proclaimed that it was time "to close the book on infectious diseases." However, by 1946 there were reports that some pathogens (such as *Staphyloccus aureus*, which colonizes our nasal passages) had become immune to its effects. The evolution of antibiotic resistance forced physicians to begin treating *S. aureus* infections with other antibiotics such as methicillin. However, this favored the few *S. aureus* that were naturally resistant to methicillin. Their descendants increased relative to nonresistant *S. aureus*, thereby forcing physicians once again to find new antibiotics to treat *S. aureus* infections. One of the most recent antibiotics to be used for these infections is vanomycin, but several strains of *S. aureus* already resist its effects. In many Asian countries, 70–80 percent of the bacteria isolated from diseased tissue include antibiotic-resistant strains of *S. aureus*. In the United States, the figure is about 40 percent. These changes in the bacterial populations and the accompanying consequences (e.g., today, *S. aureus* infections cost $30 billion per year) are predicted by Darwin's theory.

This scenario—namely, the evolution of bacteria that are immune to the effects of a drug—has occurred countless times, and has produced widespread resistance to many antibiotics. For example,

Sulfonamides, a large group of antibiotics used to treat several types of bacterial infections, were introduced in the 1930s. Resistance to sulfonamides appeared less than a decade later.

An 11-year study in Switzerland found that no strains of *Escherichia coli* (a common intestinal bacterium that can be pathogenic) were immune to any of the five fluoroquinolone antibiotics. When the percentage of patients receiving these antibiotics increased from 1.4 percent to 45 percent during the following 3 years, 28 percent of the strains became resistant to all five of the antibiotics.

A study in Atlanta found that 25 percent of patients had *Streptococcus pneumonia* that resisted penicillin. More than 25 percent of the strains of these pathogens were also resistant to several other antibiotics.

Neisseria gonorrhaeae is the bacterium that causes gonorrhea. Its resistance to penicillin tripled during the 1980s. Today, penicillin is virtually useless to combat the pathogen.

In 1941, pneumococcal pneumonia could be cured by administering 10,000 units of penicillin four times per day for 4 days. Today, giving patients 25 *million* units of penicillin every day has no effect.

In 2003, a study published in the *New England Journal of Medicine* reported that 5–10 percent of patients admitted to hospitals acquired a bacterial infection during their stay. The probability that such an infection will resist at least one antibiotic has steadily risen during the past few decades.

Today, almost half of all infections by *Staphylococcus* in hospitals are immune to most antibiotics. On a larger scale, more than 70 percent of the strains of bacteria that cause infections in hospitals are resistant to at least one of the drugs that was previously used to treat the infection.

In 1994, the Food and Drug Administration approved the use of antibiotics called quinolones to prevent infections of chickens by the intestinal bacterium *Campylobacter jejuni*. Since then, quinolone-resistant *Campylobacter* strains in humans have risen from 1 to 19 percent.

Beginning in 1978, physicians in Finland began treating middle-ear infections (caused by *Moraxella catarrhalis*) with penicillin-like antibiotics. Although these antibiotics initially cured virtually all of these infections, by 1993 more than 90 percent of the infections were immune to the antibiotics.

More than 2,000,000 people acquire an infection each year in U.S. hospitals. More than 90,000 of these patients die as a result of their infection (in 1992, only about 13,000 died from infections incurred at hospitals). Most of these infections are resistant to more than one antibiotic that was formerly used to treat the infection. Almost every hospital in the world treats people with resistant bacteria in our battle with these ever-changing pathogens. Ironically, and thanks to evolution, our attempts to vanquish these agents of disease have unintentionally made them stronger.

Note here that the antibiotics did not cause resistance to appear; the gene for antibiotic resistance arose spontaneously, and the presence of antibiotics merely favored the survival of bacteria with resistance. Also note that the gene for antibiotic resistance is neither "good" nor "bad"—the resistant bacteria are favored only because the antibiotics are present.

Not surprisingly, organisms have evolved resistance to other drugs, as well as to pesticides (see Pesticide Resistance, Chapter 3). For example, HIV has very high rates of mutation, and drug resistance can evolve in HIV in just a few days. As a result, each patient is infected with a unique version of the virus. Reflecting the finch evolution that occurred on the Galápagos Islands, each person infected with HIV represents a separate island inhabited by a unique group of viruses.

If antibiotics select for resistance, then the level of bacterial resistance to antibiotics should track the consumption of antibiotics. A test of this prediction shows that this is what happens; countries with the highest rates of antibiotic consumption per capita are those where the highest rates of antibiotic resistance usually occur.

Antibiotic resistance is driven by our widespread use of antibiotics. Indeed, farmers alone use more than 25 million pounds of antibiotics per year. The consequences are just as Darwin's theory predicts: Antibiotic resistance is why diseases such as tuberculosis, malaria, gonorrhea, and childhood ear infections are more difficult to treat than they were just a few decades ago. Although more than 100 antibiotics are used today in clinics, drug companies spend millions of dollars each year to discover new antibiotics to replace ones that have been made ineffective by the evolution of antibiotic resistance. Even if you never get sick, drug resistance will affect you, because it increases the cost of health care in the United States alone by billions of dollars every year.

Moreover, evolution tells us that drug resistance may have been lurking in the bacterial genome long before it was encouraged to increase by our use of antibiotics. It is no accident that we derive many of our antibiotics from fungi. Fungi have been at war with their competitors (including bacteria and other microbes) for countless ages. In other words, bacteria have seen these kinds of compounds long before humans ever donned the first white coat, or picked up the first stethoscope. We are simply the latest players to join an unimaginably long-running contest.

CAN EVOLUTION BE STOPPED?

Despite the overwhelming amount of evidence for evolution, many people continue to claim that evolution does not occur. For this to be true—that is, for there to be no changes in gene frequencies in a population—five conditions must be met:

1. *No mutations.* There can be no mutations.
2. *No gene flow.* Alleles cannot migrate into or out of the population.
3. *No genetic drift.* The population must be large, and changes in gene frequencies due to chance alone must be insignificant.
4. *Random mating.* Individuals mate by chance and not according to their genotypes or phenotypes.
5. *No selection.* There can be no selective agents favoring one genotype over another.

These conditions are never met, even in artificial conditions—for example, mutations cannot be stopped from occurring. Clearly, then, the question is not whether evolution occurs; *it does*. Moreover, it cannot be stopped from occurring.

CONTRADICTORY EVIDENCE?

All scientific theories must be falsifiable; that is, it must be possible to observe phenomena or to perform experiments that would cause scientists to conclude that the theory is invalid. There are many possible observations that would invalidate evolution. For example, the first primates evolved about 60 million years ago, whereas trilobites had lived in ancient oceans from 500 to 245 million years ago, when they became extinct. If fossils of trilobites and humans were found side by side in a single fossil bed, scientists would have to seriously modify (or even reject) many of their claims about evolution. Similarly, evolution would have to be discarded if scientists were to convincingly document human fossils alongside dinosaur fossils. However, no such "smoking gun" evidence has been found. Competent biologists debate details about evolution such as rates and specific common ancestors, but they do not question whether evolution occurs. It does. Evolution has been affirmed by *thousands* of scientific studies. Scientists do not publish papers announcing "new evidence for evolution" anymore than chemists publish articles showing that water is made of oxygen and hydrogen. Nevertheless, many people continue to fear and oppose evolution, and often voice their grievances in court (see Evolution in the Courtroom).

Today, biologists continue to test evolutionary ideas, and these tests have produced a range of intriguing questions. For example, when did various species first evolve? What is the pace of evolution? How much is the evolution of one species related to the evolution of another? These and other questions continue to help us understand life's history on Earth. For example, a hypothesis known as *punctuated equilibrium* has generated much interest; this hypothesis, proposed by Niles Eldredge and Stephen Jay Gould in 1972, states that a species that has been stable for long periods can evolve into new lineages in a period as short as a few thousand years. Many biologists continue to test this and other hypotheses and, in the process, help us learn more about evolution and the history of life on Earth.

Although debates about these and other evolutionary questions are often interesting, it is important to note that these questions are dealing with various aspects of evolution, not with whether evolution occurs.

Indeed, evolution itself has been tested and confirmed by thousands of carefully controlled scientific studies in disciplines ranging from paleontology and biogeography to comparative anatomy and molecular biology. Moreover, there is no well-documented scientific evidence that contradicts the occurrence of evolution by natural selection. As Richard Lewontin noted in 1981, "Evolution is a fact.... Birds arose from nonbirds, and humans from nonhumans. No person who pretends to any understanding of the natural world can deny these facts any more than she or he can deny that the Earth is round, rotates on its axis, and revolves around the sun." The overwhelming evidence for evolution even convinced Pope John Paul II, who told the Pontifical Academy of Sciences in 1996 that "It is indeed remarkable that [Darwin's] theory has been progressively accepted by researchers, following a series of discoveries in various fields of knowledge. The convergence, neither sought nor fabricated, of results of work that was conducted independently is in itself a significant argument in favor of [Darwin's] theory."

EVOLUTION IN THE COURTROOM

The lack of a scientifically valid alternative to Darwin's theory of natural selection has not stopped antievolutionists from trying to forbid or undermine the teaching of evolution. The so-called evolution–creationism controversy began in the United States in the 1920s when three Southern states (i.e., Tennessee, Mississippi, and Arkansas) passed laws banning the teaching of human evolution. In 1925, Tennessee's antievolution law produced the spectacular (but legally meaningless) Scopes "Monkey Trial." (Substitute science-teacher John Scopes was convicted of the misdemeanor of teaching human evolution, but his conviction was later set aside by the Tennessee Supreme Court.) In the 1960s, all of the laws banning the teaching of human evolution were declared unconstitutional by the U.S. Supreme Court's *Epperson v. Arkansas* decision, after which antievolutionists declared that the Bible is a science book that deserves "equal time" and "balanced treatment" with evolution in science classrooms. When courts ruled in the 1980s that so-called "creation science" is not science and that its teaching is unconstitutional, the antievolutionists began demanding that evolution be balanced by "intelligent design," a repackaging of Paley's claim that life is so complex that it must have come from an intelligent designer (Chapter 1). Late in 2005, Federal Judge John Jones ruled that "intelligent design" is religion, not science.

Antievolutionists have lost every legal challenge to the teaching of evolution in public schools. A summary of these decisions is provided in Appendix 2.

The Scopes Trial accomplished nothing from a legal perspective, yet remains the most famous event in the history of the evolution–creationism controversy. The trial occurred in Dayton, Tennessee, in the summer of 1925, and is referenced in virtually all newspaper and television stories about the teaching of evolution. (*Randy Moore*)

SUMMARY

Evolution is a continuing process that explains and accurately predicts the history of life on Earth, the unity of life, and the diversity and variation of species. Evolution is supported by a vast amount of evidence, including fossils, extinctions, artificial selection, embryology, vestigial structures, biogeography, comparative anatomy, and molecular biology. The explanatory power of Darwin's idea, combined with scientists' inability to reconcile this evidence with any other explanations for life's diversity, is the basis for the now-famous quotation by biologist Theodosius Dobzhansky that "nothing in biology makes sense except in the light of evolution."

3

HOW EVOLUTION WORKS

In the previous chapters, you have seen evidence for evolution, and you have learned how the concept developed. We will now consider how evolution works on a daily basis. After all, although evolution explains how horses "lost" their toes and gained hooves and why the cilia in your air passages are identical to the cilia of a flatworm or a snail—in other words, although evolution explains how the animals and plants and bacteria and viruses of the world came to be the way they are—evolution doesn't stop there. It continues as a vital process in the world today, giving us antibiotic-resistant pathogens (Chapter 2) and insects that defeat our best chemical controls (see Pesticide Resistance). How does it do this? How does evolution work? If evolution is, as Darwin noted, "descent with modification from a common ancestor," what is modified, and how?

GENES LINK GENERATIONS

To understand what evolution modifies, and how, we must first think about genes. *Genes* link generations by transferring instructions for how to develop and operate an organism from the parent to its offspring. Genes are made of DNA (deoxyribonucleic acid), a molecule that is passed in eggs and sperm from one generation to the next. Genes can be seen as biochemical information that is translated into traits such as hair color and blood type. In other words, the traits that an organism inherits are based on information in its genes. The more genes you share with other descendants of an ancestor, the more inherited traits you will have in common with those descendants (see DNA and Chromosomes).

〰〰

PESTICIDE RESISTANCE

From observing our skyscrapers and our computers, you might think that we humans rule the planet, without competition. You would be wrong. Insects can be tremendously beneficial. They pollinate 80 percent of flowering plants, they act as biological control agents, and they even produce honey. However, they also consume approximately one-third of our agricultural output and transmit malaria (over 200 million cases), West Nile virus, Dengue Fever, and other diseases. Because of this, we have tried to control their numbers, if not exterminate them, using chemical poisons. The results are exactly what you would predict, having learned about evolution: insects that have the genetic ability to resist the ill effects of pesticides are favored and leave more offspring than other insects do.

If you return to the discussion in Chapter 2 (see "Evolution, Life, and Death: The Evolution of Drug Resistance") about how natural selection works, you can see how pesticide resistance evolves. Not all individuals of any given species of insect pest are alike, and some have genes that allow them to withstand the effects of the poison; we call these individuals "resistant." Although there may be very few resistant insects when a poison is first applied, they have a tremendous selective advantage; they live and produce offspring while most of *their* competitors are killed. These offspring inherit the resistance gene, and soon—thanks to the reduced amount of competition and their rapid rates of reproduction—the insect population returns to prepoison levels, with one annoying difference: the new population of pests carries a gene that makes them resistant to that insecticide. (There are many ways to be resistant, ranging from biochemical differences that interfere with pesticide activity to behavioral differences that reduce exposure to pesticides.) As a result, farmers must periodically change pesticides if they hope to control their insect competitors.

To appreciate all of this, consider the pesticide dichloro-diphenyl-trichloroethane, a poison better known as DDT. DDT was invented in 1939. Cheap, stable, effective, and easy to make, DDT was soon considered the ultimate pesticide. Between 1941 and 1976, almost 5 million tons of DDT were released into the environment, and people began to believe that pests could not only be controlled, but eliminated. This belief was common among people trying to control malaria, a disease that kills more than 2 million people per year.

The parasite that causes malaria is transmitted to humans by mosquitoes, and applications of DDT were used to control mosquito populations. For a while, this worked; populations of mosquitoes plummeted after being sprayed with DDT. This prompted Paul Russell of the Rockefeller University to write in his 1955 book, *Man's Mastery of Malaria*, that "for the first time it is economically feasible for nations, however underdeveloped and whatever the climate, to banish malaria completely from their borders."

But as Darwin's theory predicts, the presence of DDT should select for organisms resistant to DDT, and this is exactly what happened. Although the applications of DDT saved lives and crops, by 1950 scientists had discovered some insects that were resistant to DDT. By 1992, more than 500 species resisted DDT. Today, that number continues to grow. In the case of malaria, the evolution of pesticide-resistant mosquito vectors has converged with the evolution of drug-resistant malaria organisms.

Although the use of DDT was banned in the United States in 1972, we continue to depend on other pesticides to produce crops large enough to feed our growing population. Indeed, every year, Americans use more than 2 million tons of pesticides; this is more than 20 times the amount we used in 1945, and these pesticides are up to 100 times more toxic than their predecessors. Nevertheless, the fraction of crops lost to insects continues to climb.

Over 500 species of insects, spiders, and mites are now resistant to the poisons we have used on them. At least 17 of those species are resistant to multiple poisons. Our quest to eradicate pests with pesticides such as DDT has failed because, whether we like it or not, evolution by natural selection happens. Ignoring it will not make it go away—it will simply produce more resistant strains and species. The evolution of resistance—to pesticides, to antibiotics—presents us with an irony: The failure to take evolution into account has proven to be perhaps the major defeat of 20th century science.

DNA and Chromosomes

DNA molecules are long strings of chemicals; you might imagine them to be strands of beads. The order of the different "beads" is like a code that tells a cell what to do and what kinds of structures to make. DNA is found in, and behaves the same way in, all organisms; this is a unifying trait of life. It is this information that is passed on from parent to offspring. (Some viruses use a closely related molecule, RNA, in a similar manner).

Chromosomes are coiled batches of DNA. The cells that form gametes such as eggs and sperm divide in a way that gives each gamete one half of the chromosomes found in a normal, nongamete cell. When these gametes unite during fertilization, the resulting embryo has two sets of chromosomes, one set inherited from each parent. As the embryo grows and becomes an adult, its cells divide; when they divide, the chromosomes reproduce themselves as well. Because of this, every cell in the body (except the gametes) has the same two sets of chromosomes.

A *gene* is a certain portion of DNA (i.e., chromosome) that contains the code for a specific trait. Thus, an organism has two genes for each trait—one gene from each parent. If all organisms in a population have genes that specify the same information for a given trait, there is no genetic variation for that trait.

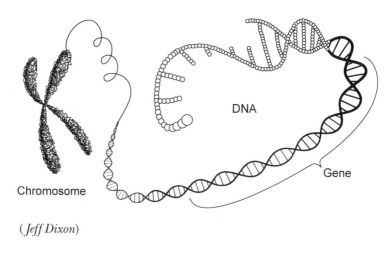

DNA

Gene

Chromosome

(*Jeff Dixon*)

Of course, not all traits are genetic. Some result from environmental influences. If you spend every waking moment on the beach, your naturally brown hair may look blonde and sun-bleached. That "blonde" trait is not genetic—it is an acquired, environmentally induced trait, and will not be passed on to your offspring. (To have that kind of blonde hair, they will have to join you on the beach.) When we think about evolution, we do not think about Lamarck's "acquired traits" (Chapter 1), but instead about genetic traits—that is, traits that are inherited.

THE FREQUENCIES OF GENES CAN CHANGE

Over time, the *frequency* (relative number) of genes for a given trait in a population of organisms can change. Imagine a population of wafflesmoozers, most of them white. (A wafflesmoozer is a little known animal that is so secretive there is no evidence of its existence; see Figure 3.1.) If you return to this population in 20 wafflesmoozer generations, you might find that the descendants are more likely to be black wafflesmoozers than white wafflesmoozers. If wafflesmoozer color is genetically determined, then the frequencies of genes for black, as well as genes for white, have changed over time. That is, the gene frequencies have been modified. If other traits such as whisker length, tail curl, ear size, and the like are also under genetic control, and if the frequencies of those genes change as well, resulting in bigger ears, curlier tails, and longer whiskers, then eventually the big-eared, curly-tailed, long-whiskered, black descendants of the original population that was mostly small-eared, straight-tailed, short-whiskered, white wafflesmoozers will look very different from the ancestors. This is descent with modification. The population changed over time.

That seems straightforward enough, but how in the world did a bunch of small-eared, straight-tailed, short-whiskered, white wafflesmoozers give rise to such different descendants? How did these traits get modified?

WHAT CAUSES GENETIC VARIATION?

Reexamine the summary of natural selection in Chapter 2. The first thing that was necessary for modification of wafflesmoozer gene frequencies to occur was *genetic variation.* That is, for evolution to occur within the population of wafflesmoozers, the genes that transmitted those inherited traits couldn't all be alike. If all of the wafflesmoozers that were first observed were purebred for small ears, straight tails, short whiskers, and white coat color—in other words, if there were no genes in the population for anything but those traits, there would be no variation, and it would be very likely that no matter how many wafflesmoozer-generations passed, and no matter what environmental changes occurred, the descendant wafflesmoozers would look like the ancestors. When we think of gene frequencies changing, we usually think of a common gene (in our example, a gene for white coat color)

Figure 3.1 Meet the wafflesmoozers (whimsical animals). One is white, and has two white genes; the other is white, and has a white gene and a black one. (The genes are not literally white or black, but they code for those colors.) As you can see, a population of white wafflesmoozers might be genetically variable with reference to color. (*Jeff Dixon*)

becoming less common; in that case, some other coat color—resulting from another gene—becomes more common. For that to happen, the other gene must exist in the population. This is called *genetic variation*. If there are no genes for anything other than white coat color, then there is no genetic variation for that trait.

You may wonder how the population could be all white if genetic variation existed. "Dominant" genes are genes that are always expressed; "recessive" genes can only be expressed in the absence of the dominant gene—that is, when paired with another recessive gene. If the gene that specified white color was dominant to the one that specified black, and if there were only a few black genes in the population, all occurring with white ones, then there could be genetic variation in a population of all white individuals (Figure 3.1).

Genetic variation itself comes from three major sources: mutation, the movement of organisms (and their genes) into an area, and the mixing of genetic material that occurs during sexual reproduction. Let's consider mutation first.

Mutation Can Cause Genetic Variation

A *mutation* is a change in an organism's DNA, that is, a change in the information that governs inherited traits. So let's say that our wafflesmoozers were indeed purebred for small ears, straight tails, and all the rest. Even under such circumstances, if a mutation occurred in the eggs or sperm of a wafflesmoozer—say, a mutation that made the wafflesmoozer's tail curl—the offspring of that wafflesmoozer would carry different information than the offspring of the other wafflesmoozers, and *Voila!* there would be genetic variation. Many genes mutate 1 to 10 times per 100,000 cell divisions. However, no matter how rare, mutation is the source of variation in a population. Mutations provide the raw material for evolutionary change.

You may have noticed that the mutation with which we are concerned occurred in the gametes (eggs or sperm) of the wafflesmoozer. What about mutations that occur elsewhere? What about mutations that arise, say, in the cells on the tip of the nose? It's true that mutations can occur in any DNA in any cell of the body, but only the mutations that occur in gametes (or cells destined to become gametes) are passed on to new generations and contribute to genetic variation. What happens in cells on the tip of the nose stays there.

There is one other important thing to remember about mutations— they are random. No matter how much you may "need" one, you cannot make a certain kind of mutation happen; new genes do not arise on demand, and an organism cannot order the traits it needs from some on-line catalog of genetic goodies. The rate of mutation can be influenced by environmental insults like radiation or some chemicals (especially carcinogens), and some areas of DNA are more likely to mutate than others. None of this alters the fact that mutations have random effects; some are beneficial, many are detrimental, and some matter hardly at all. After all, a random change to a well-functioning organism is more likely to disturb that function than improve it.

Gene Flow Can Cause Genetic Variation

Another way that a population of wafflesmoozers that had no genetic variation (i.e., was purebred for small ears, straight tails, short whiskers, and white coat color) could eventually change over time and evolve into some other morph would be if other wafflesmoozers that carried genes for other traits migrated into that population. This is called *gene flow.* When that happens, the resulting population (natives and immigrants) does have genetic variation, even it if didn't before the arrival of the immigrants.

Genetic Mixing Can Cause Genetic Variation

Finally, genetic variation can result from *genetic mixing* that occurs during sexual reproduction. Chromosomes are compact strands of DNA; we get one set of chromosomes from each of our parents, and that combination of information produces us—our parents' offspring. These chromosomes are reproduced in every cell in our bodies and, because they contain our DNA, they contain the information that specifies our inherited traits. When any of us produces gametes, those two sets of chromosomes line up next to each other. During this process, a chunk of the information that you received from your mother may switch places with a chunk of similar information from your father. In other words, the piece of the chromosome that carries, say, information about hair color from your mother may switch places with the piece of the chromosome from your father that also carries information about hair color. Thus, each chromosome remains functional—the information about hair color in general did not disappear—but the information that you carry about your dad's hair color is now sitting in the middle of a chromosome full of other information that you inherited from your mother. When this happens, combinations of genetic traits get shuffled, and that increases genetic variation. (If you wish to learn more about this and other basic genetic processes, see Appendix 4.) Note that of the three major sources of genetic variation—mutation, gene flow, and genetic mixing—only mutation actually generates new genetic material.

EVOLUTION IS CHANGE IN GENE FREQUENCY—HOW DO GENE FREQUENCIES CHANGE?

We can now answer the first of our questions: If evolution is "descent with modification from a common ancestor," what is modified? In the course of evolution, *gene frequencies in a population are modified (that is, they change) over time*, and as a result, inherited traits in that population change. Note that evolution is genetic change in a population, not genetic change in an individual—this is why populations evolve, and individuals do not. We see an increase in big-eared, curly-tailed, long-whiskered black wafflesmoozers. That brings us to our second question: How does this happen? Scientists are still discovering many things about how evolution works, but four mechanisms are widely accepted: mutation, migration, genetic drift, and natural selection.

We've already seen one of these mechanisms of evolution—*mutation.* Not only does mutation provide the genetic variation necessary for evolution, but when mutation occurs, gene frequencies also change, and technically, that means that evolution happens. Mutation is occurring all the time, but it occurs at very low rates, and as we indicated previously, it is random. Mutation alone is not likely to produce change in any particular direction (Figure 3.2a).

Genetic drift, like mutation, is an accidental source of altered gene frequencies. Sometimes, by chance—for example, by getting drowned in a hurricane, getting squashed by a falling tree, or getting uprooted by a bulldozer—some individuals produce fewer offspring than others. This is not because they are better adapted to the environment, but because they are simply unlucky. This uneven distribution of ill fortune nonetheless produces changes in the genetic composition of a population and thus causes evolution (Figure 3.2b).

As you might imagine, the process of genetic drift is especially influential in small populations, where a few drowned, squashed, or otherwise obliterated individuals can have a large effect on future gene frequencies; some kinds of genes may disappear entirely. Unlike mutation or migration (see below), genetic drift reduces genetic variation. Once such reduction occurs, it takes a long time for mutation and migration to generate additional variation.

The reduction of genetic variation as a result of genetic drift is very important in conservation biology, where two similar phenomena—bottlenecks and founder effects—can mar the recovery of a rare species. During a *bottleneck*, the population is greatly diminished and thus vulnerable to genetic drift. Even if the bottleneck lasts for only one or two generations before the number of individuals increases again, if genes are lost in the process, then the "recovered" population will be less genetically diverse. A *founder effect* occurs when a new population is started by a few organisms. Again, a reduction in genetic diversity is a probable outcome, because it is unlikely that all of the genes from the original population have made it into the new one. The founder effect is especially important in the evolution of organisms that live on oceanic islands, such as the Galápagos Islands, which Darwin visited. Most of the kinds of organisms that live on these islands were probably derived from one or a few founders.

Zoos are among the organizations at the forefront of conservation biology, and founder effects are one of their major challenges. Scientists in charge of captive breeding programs study pedigrees carefully and attempt to maximize genetic diversity. Similar attention is directed at recovery programs in the field (e.g., whooping cranes, Florida panther; see Conservation and Genetic Diversity.)

Figure 3.2 Mechanisms of evolution. (a) **Mutation** occurring in a wafflesmoozer, producing changes in gene frequencies; in this case, a mutation for long ears has occurred; (b) **Genetic drift** in wafflesmoozers, producing changes in gene frequencies; in this case, two wafflesmoozers have been struck by lightning. They just happen to both be white, small-eared wafflesmoozers, and as a result, the proportion of such wafflesmoozers in the population is greatly diminished; (c) **Gene flow** in wafflesmoozers, producing changes in gene frequencies; (d) **Natural selection** producing changes in wafflesmoozer gene frequencies; in this case, two dark, long-eared wafflesmoozers are snatched up by hungry Ear-Nabbing Ravenawfuls. Natural selection does not favor long ears and dark colors in areas where wafflesmoozers overlap with Ear-Nabbing Ravenawfuls. (*Jeff Dixon*)

Migration, or gene flow, is another source of genetic variation that can drive evolution. When organisms (and their genes) enter an area and also interbreed with residents, the next generation will exhibit different gene frequencies than if the immigrants had not arrived (Figure 3.2c).

───────────────────────── ༄ ─────────────────────────

CONSERVATION AND GENETIC DIVERSITY

Habitat fragmentation is the process of breaking up once-large stretches of habitat into small chunks that are separated by swaths of altered terrain. This produces multiple small and isolated populations, which in turn play havoc with gene flow. For example, millions of greater prairie chickens occupied most of Illinois 200 years ago; by the 1990s there were 50 individuals left in two pockets of remaining habitat. Despite a ban on hunting prairie chickens through most of the last century, the population in one of those habitats (Jasper County) crashed mysteriously after a brief recovery in the 1970s. Scientists hypothesized that small population size, genetic isolation, and genetic drift had interacted to cause the prairie chicken disaster. In a small population, drift may result in an abundance of some deleterious, or "bad," genes. If these are the only genes left to influence a given trait, then without some immigration or marvelously lucky mutation, all individuals in the population are likely to have those genes, and this will work to their disadvantage. Scientists tested this hypothesis by measuring genetic diversity in the Jasper County population, other prairie chicken populations, and the ancestors of Illinois prairie chickens, some of which can be found in museums. They found that the Jasper County prairie chickens had very low genetic diversity by comparison. To increase genetic diversity and reverse this downward spiral, scientists introduced prairie chickens from other locations. Survivorship of Jasper County prairie chickens is now increasing. This experiment does not prove that drift and genetic isolation were the causes of the decline, because it is always possible some other environmental variable could have been at work, but it does mean that drift and genetic isolation are very likely explanations. (A word of caution here: introductions of new stock like that of the prairie chickens must be performed very carefully, lest disease or other unintended challenges be introduced as well.)

───────────────────────── ༄ ─────────────────────────

Although mutation, genetic drift, and migration all produce evolutionary change, none of these mechanisms necessarily results in traits that help the organism survive and reproduce. Mutation and genetic drift are random events, and migration is simply the movement of individuals into a population; there is no reason to expect the immigrants to survive or reproduce better than the residents do. The fourth mechanism of evolution—*natural selection*—is the only one that increases the

frequency of inherited traits that, in turn, increase survival and reproduction. Such traits are called *adaptations*. How does natural selection
do this?

Natural Selection Produces Adaptive Change in Gene Frequencies

First of all, there must be variation in inherited traits, in other words,
genetic variation. Recall our wafflesmoozers from the beginning of this
chapter. Although most of the individuals in the initial population of
wafflesmoozers were small-eared, straight-tailed, short-whiskered, and
white, there had to be some genetic variation for the population to
become big-eared, curly-tailed, long-whiskered, and black.

Second, not all individuals reproduce with equal success. Some may
get eaten, and others may not be as good at acquiring resources. For instance, if a wafflesmoozer with small ears, a straight tail, short whiskers,
and/or white fur is likely to be pounced upon by a cat or unlikely
to find food, and if a wafflesmoozer with big ears, a curly tail, long
whiskers, and/or black fur frustrates cats and finds the best cheese, then
the latter wafflesmoozer is more likely to survive and leave offspring.
For that matter, perhaps big-eared, curly-tailed, long-whiskered, black
wafflesmoozers are simply more fertile than their small-eared counterparts. Any of these possibilities means that in the next generation of
wafflesmoozers, the genes for large ears, curly tails, long whiskers,
and/or black fur will increase relative to other genes. In other words,
traits that give the wafflesmoozers an advantage and that help them
survive and reproduce will increase in the population.

Although the causes of genetic variation are accidental and nondi-
rectional, natural selection itself is anything but accidental and nondi-
rectional. Unlike evolution that results from mutation, migration, and
genetic drift, evolution by natural selection results not only in changed
gene frequencies, but also in the kinds of changes that increase adaptations. Organisms that have traits that produce many surviving and
reproducing offspring (which also have those traits) will increase as
time goes on.

In a nutshell, this is what "fitness" is all about. We hear a lot about
"survival of the fittest," and news reporters would have us believe that
the "fittest" are those who win races, persist in "reality" shows, take over
corporations, and generally get what they want. In the world of biology, however, fitness means something quite different. *Fitness* has two
components: survival and reproduction. It is about nothing more or less
than the ability to contribute genes to the next generation. We know, for
instance, that if in the process of vigorously defending a territory, a male

bird does not feed his offspring well enough, he will have fewer surviving chicks than his less vigilant and less quarrelsome neighbor. No matter how much the more boisterous male sings, displays his bright feathers, and pecks at intruders, if all of this means that his chicks don't survive as well as they would have with less singing and pecking, then he is less fit than a bird with more surviving chicks. Charles and Emma Darwin, who produced three children that became parents themselves, had a relatively high level of fitness. George Washington, on the other hand, never had any children. He may have been the father of his country, but his direct fitness was zero. (We'll discuss indirect fitness later in this chapter.)

Some folks can get fairly enthusiastic about adaptations; they try to find adaptive explanations for everything they see. As in all of science, it is useful to employ a bit of skepticism when explaining the natural world. Not every trait is an adaptation. First of all, as we have seen, not all changes in gene frequencies result from natural selection; although a population may contain several versions of a gene, it may be that none of the versions affects fitness. Such genes are thought to be "neutral," and while their frequencies may change as a result of drift or similar processes, they are not subject to natural selection. An adaptation, on the other hand, results from natural selection. Likewise, some traits may have lost their previous adaptive function—recall the pelvic bones of whales, the vestigial traits discussed in Chapter 2. These traits may have been adaptive at one time, but are no longer adaptive. Finally, some traits may not be adaptive themselves, but may be byproducts or constraints of other adaptations. For instance, one might argue that natural selection should favor organisms that produce a near-infinite number of robust healthy offspring. If resources are limiting, however, this proves to be impossible. Instead, depending on the environment and the likelihood of offspring survival, organisms may enhance their fitness by producing very few offspring in which they invest heavily, or many offspring in which they invest little.

Although all adaptations increase survival and reproduction, not every trait that does so is an adaptation—that is, not every trait that increases survival and reproduction is a product of natural selection. This is true because not all traits are inherited. The fact that you don't cross the street in front of a speeding bus undoubtedly increases your chances of survival and reproduction, but that trait is not genetically based, and therefore is not an adaptation. Not only does an adaptation increase survival and reproduction, it is also an inherited trait.

Finally, an adaptation performs a current function that increases survival and reproduction. Most clams have very large gills, covered with cilia. As the cilia move, they draw water across the gill. Oxygen in the moving water is transferred across the surface of the gill into the blood that flows inside the gill, and while this is happening, food particles also carried by the water are trapped on the mucus that covers the gill. This food-laden mucus is moved (by cilia, again) to the mouth, where it is ingested. Gills are fine adaptations, the result of natural selection favoring oxygen acquisition. They are full of blood vessels and they have a large surface area for absorption. However, when it comes to food gathering, gills can be seen as exaptations. An *exaptation* is a trait that may be beneficial, but that is used for something other than its original function. Gills evolved in response to selective pressure favoring efficient oxygen acquisition and can be found in many clams, including some that do not use them in feeding. Once gills had evolved, many clams used them for food gathering as well as oxygen acquisition (Figure 3.3). Similarly, African black herons use their wings to eliminate glare on the surface of water, thereby getting a better view of the fish and tadpoles they seek for food. And in the giant panda (a zoo favorite), a wristbone has evolved into a crude "thumb" for holding and stripping the bamboo stalks that the panda eats. Clearly, these adaptations were not produced from scratch, but instead by co-opting existing structures and adapting them to new situations. Natural selection favors the fittest variation that is *available*, not that which is theoretically *possible*.

Flowering plants provide us with many examples of adaptations. Because plants cannot move around, their male gametes (which are carried in pollen) must find other ways to get to places where they can fertilize female gametes. As Darwin noted, plants such as orchids go to fantastic lengths to accomplish this. Some orchids even mimic the appearance and smell of female insects, all the better to attract male insects, which then accidentally pick up pollen (strategically placed, of course) and transfer it to other plants.

Organisms that transfer pollen for plants are called "pollinators." Pollinators can be tricked into transporting pollen, as in the case of those orchids, or they can be bribed with nectar. Yet other flowers attract such unlikely pollinators as carrion flies by producing the kinds of odors that only a carrion fly could love. These flies (and beetles) are attracted to the odoriferous flowers. There, they accidentally pick up pollen and fly away with it to the next smelly flower, where the pollen can fertilize the eggs of that plant.

We do not expect every trait of an organism to be adaptive, nor do we expect organisms to have every trait that they "need." As we have indicated, some traits are historical leftovers (recall the vestigial characters of

Chapter 2) or they may be byproducts of more adaptive traits. Moreover, natural selection can only act on variation that is present. Pigs might find wings very useful, but for a variety of developmental and physical reasons, the winged version of a pig has not presented itself for natural selection ... and if it did, there is some question about how big a breast bone would be needed to attach the huge muscles that would power the wings lifting the porcine wonder off the ground.

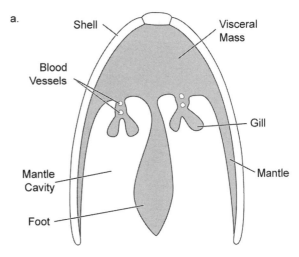

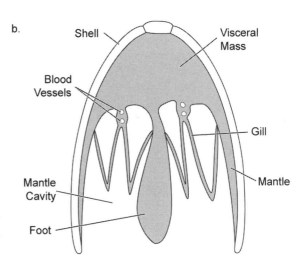

In summary, adaptations are inherited, they result from natural selection, and they have evolved to perform a current function that increases fitness. In fact, there is increasing evidence that sexual reproduction itself is an adaptation. That may sound a little strange at first, but recall that many organisms in the world actually clone themselves—that is, they do not depend solely upon sexual reproduction, but simply divide and produce clones of themselves. Plants are very good at asexual reproduction, as you know if you have ever grown day lilies or strawberries; these plants and others like them can send other plants up from underground stems or roots.

Figure 3.3 The clam gill was originally used for oxygen exchange (a), but later modified and enlarged as its use in filter feeding was favored (b). (*Jeff Dixon*)

We exploit this trait when we propagate plants from their pieces (i.e., "cuttings"). Animals such as corals produce new individuals as offshoots from the bodies of existing individuals; this is how a coral reef grows.

Sexual   Asexual

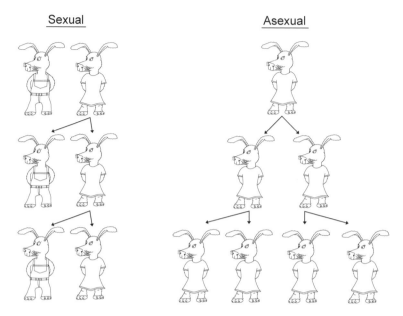

Figure 3.4   Genetically, it costs twice as much to reproduce
sexually—that is, to produce sons and daughters—as to reproduce
asexually and produce only daughters. (*Jeff Dixon*) Modification of
Figure 1, page 1023, *Encyclopedia of Evolution* (Pagel, ed.).

Fragments of many worms can grow into new individuals, and the eggs
of some fish and insects can hatch into new individuals without ever
being fertilized. If you consider the concept of fitness, you will realize
that if everything else is equal, such organisms are indeed fit; each
of their offspring contains 100 percent of the parent's genes. That's
quite an accomplishment when it comes to passing on genes. On the
other hand, organisms that reproduce sexually contribute only 50
percent of their genes to each offspring; half of the offspring's genetic
material comes from one parent, and half from the other (Figure
3.4). Thus, a sexually reproducing organism has to produce twice as
many offspring as an asexually reproducing one if it is to pass on as
many genes. If fitness is indeed about passing on genes, then sexual
reproduction probably shouldn't exist. How, then, can we suggest it is an
adaptation?

Sexual Reproduction Increases Genetic Variation

Recall that during sexual reproduction, genetic mixing can occur
at the chromosome level, with some chunks of chromosomes trading

places as eggs and sperm are formed. In addition to this mixing that occurs during gamete formation, fertilization offers yet additional opportunity for novel combinations of genes. When an egg from one individual is fertilized by the sperm of another, the resulting embryo is a unique organism, containing some genetic information from each parent. In the absence of twinning (when one fertilized egg forms two embryos), each offspring from two parents will differ from the parents, from every other offspring of those parents, and from every other member of the species.

Sexually reproducing plants and animals may have to produce twice as many offspring as asexually reproducing parents do to transfer the same amount of genes to the next generation, but the variation in sexually produced offspring increases the likelihood that at least some of them will find a way to survive and reproduce in a changing environment. Asexually produced offspring, on the other hand, are identical to the parent and to each other. If conditions remain stable, there is a good chance that asexually produced organisms will survive and reproduce, as their parent did. The earth, however, is not well known for being particularly stable. In a changing environment populated by pathogens and predators, drenched in rain and blistered by sunlight, natural selection favors sexual reproduction. Parasites and pathogens, which themselves evolve, are especially challenging selective forces that favor sexual reproduction. Put another way, there is no point in buying ten identical lottery tickets; if you are going to buy ten, you will be ten times more likely to win if the tickets are all different. The advantages of this diversity are so great that even organisms with the ability to reproduce asexually often engage in sexual reproduction as well (see The Red Queen and Sexual Reproduction).

THE RED QUEEN AND SEXUAL REPRODUCTION

The Red Queen Hypothesis states that parasites are the primary selective force favoring sexual reproduction. In Chapter 2 of *Through the Looking-Glass* (Lewis Carroll's 1872 sequel to *Alice's Adventures in Wonderland*), Alice decides to leave the looking-glass house to see the garden. Alice departs on a path that appears to be straight, but she soon discovers that the path leads her back to the house. When she speeds up along the path, she returns to the house faster. Alice then meets the huffy Red Queen, and they start to run faster and faster. Alice, however, is perplexed, because neither she nor the Red Queen seems to be moving, and when they both stop running they are in the same place. The Red Queen then tells Alice that "Now, *here*, you see, it takes all the running you can do, to keep in the same place."

According to the Red Queen Hypothesis, sexual reproduction persists because it enables many species to rapidly evolve new genetic defenses against parasites that attempt to live off them. That is, a constantly changing environment causes continuing evolution; populations must keep evolving to maintain themselves in constantly changing environments. Evolutionary biologist Graham Bell applied the scene of Alice and the Red Queen to evolutionary thinking about sexual reproduction: Parasites infect common host genotypes, and the hosts that produce different genotypes can stay ahead of the parasites—but they can never leave them behind.

Mate Choice Can Be Adaptive

Sexual reproduction itself is associated with a special kind of natural selection called *sexual selection*. Sexually selected traits are those that enhance an organism's ability to mate and reproduce. They may make an individual more attractive to the opposite sex or increase its ability to compete successfully for mates. The croaking of frogs, the brilliant, flashing color of some fish, and the spreading antlers of deer are only a few examples of what sexual selection has produced.

To someone who does not understand evolution, such traits may seem almost arbitrarily distributed. Among birds and mammals, for instance, why should these traits appear to be more common in males than in females? And why are they spectacular in some animals like peacocks, and hardly noticeable in others, like the kestrel or chickadee? Remember that in science, theories (like the theory of evolution) explain a wide range of observations and generate testable predictions. Evolution by natural selection—including sexual selection—can explain why peacocks (males) are showier than peahens (females), and can predict which animals are likely to follow the "showy mate" pattern, and which are not likely to do so (Figure 3.5). Darwin himself began to answer these questions in 1872 in his book *The Descent of Man, and Selection in Relation to Sex.*

The wonderful tail of a peacock is not there to beautify the world for humans, no matter how appealing that notion might be. The tail attracts mates for the peacock—lots of mates. A peacock does not spend time or energy helping the peahen rear young—in fact, with such an extravagant tail, he'd probably be more liability than asset. Instead, peacocks compete with one another for mates, and their impressive tails are one element of attraction. Because the peacock does not invest in his offspring beyond actual fertilization, he can afford to mate with many peahens, if he can only convince them to mate with him. The tail, in this regard, is everything. In such a species, a few males may mate with most of the females; most of the remaining males do not have a mate. Stakes are high. Indeed, if survival were all that mattered, such a lovely tail might never evolve, because it could cost quite a bit of energy to produce and, in addition, it might attract predators. Short of using such a tail to beat predators senseless, its benefit is hard to imagine, except in the world of mating. If peahens prefer showy tails, then males that have showy tails will leave more offspring than less spectacular males, and peacock tails will become increasingly gaudy with each generation; they will be limited only by genetic variation for gaudiness and natural selection itself on the costs of such extravagance.

Birds like male kestrels and chickadees cannot afford as many mates as peacocks because these males help incubate and feed the young. The peacock winner-take-all strategy (or winner-take-almost-all) does not work for males

Figure 3.5 An American kestrel. Peafowls are not monogamous birds and exhibit no male parental care; in American kestrels, both parents tend offspring. (*U.S. Fish and Wildlife Service*)

who must also work hard at rearing offspring; those males are limited in the number of offspring they can rear, and thus, the number of mates they have. Therefore, competition for kestrel and chickadee mates is not as severe as it is in the case of peacocks. In other words, based on evolutionary thinking, we can predict that we will find the biggest differences in appearance between the sexes (something we call "sexual dimorphism") in species where mating success is not distributed evenly, but where competition for mates is intense and a few individuals of the competitive sex may have most of the mates. Where mating success is distributed more evenly, males and females look very much alike.

In species where competition for mates is intense, not only does evolutionary thinking tell us to expect sexual dimorphism, but we also expect a certain "choosiness" on the part of the sex that is being competed for. As we have seen with the peacock and peahen, investment in individual offspring is not necessarily equal. In that case, the peahen has fewer offspring than a highly successful peacock (although more offspring than a reject peacock); she produces few eggs relative to his sperm and invests more in each one. The prediction is that the peahen that chooses her mate carefully and maximizes that investment will be favored by natural selection: her sons, like the father she chooses, will have the ability to produce beautiful tails and compete well for females; her daughters will be choosey as well.

Thus, we have two kinds of sexual selection: *intrasexual selection*—selection within a sex—occurs when male elephant seals fight for territories that will contain harems of females. *Intersexual selection*—selection imposed by one sex on the other—is selection like that imposed by peahens as they choose males such as our lovely peacock. These two kinds of sexual selection may interact, as when females prefer the winner of some intrasexual contest.

Are males always competitive, and are females always choosey? Because of the relative costs of producing an egg and a sperm, this is frequently the case. Often, male reproduction is limited by access to females, while female reproduction is limited by male quality and resources. The exceptions to this "rule" are intriguing. Our evolutionary prediction, if you recall, is that the choosey sex, like our peahen, should be the one with the greatest investment in the offspring. This is often, but not inevitably, the female. In the case of mormon crickets, for instance, the male produces a huge container for his sperm (a spermatophore) that he gives to a female, along with the sperm itself. This sperm container may be up to 25 percent of the male's mass, and females that acquire these nutrients use them to provision their eggs. In this case, the male's investment exceeds that of the female, and as predicted, he is the choosey sex; male mormon crickets with large spermatophores are courted by numerous females, and are selective about their choice of mates.

Selection for One Extreme, Both Extremes, or the Middle

The tail of the peacock is an example of *directional selection*—fitness consistently increases (or decreases) with the increase (or decrease) of a trait. Selection, if it favors intermediate forms, can also be stabilizing; in the case of *stabilizing selection* the extremes of a trait are selected against. For example, there is a fly that forms a gall in the stem of a goldenrod plant. (A "gall" is a swollen area of plant tissue that surrounds and is induced by a developing insect larva. In the case of this fly, gall size is under at least some genetic control.) As you might imagine, a small gall is little protection against some threats to the larva; parasitoid wasps, for instance, are more likely to successfully attack larvae in small galls than in larger ones. On the other hand, if the gall becomes too large, birds notice it, break it open, and eat the larva. Surviving larvae are more likely to come from intermediate-sized galls. The size of infant humans has also been influenced by stabilizing selection; infants that are too tiny have difficulty surviving, whereas infants that are too large present difficulty at birth.

Disruptive selection is the opposite of stabilizing selection. In disruptive selection, extreme forms of a trait are favored, and intermediate forms are selected against. It is likely that the difference between eggs and sperm may have resulted from disruptive selection. Even if gametes were once the same size, it would only take a slight amount of variation in size to begin the disruptive selection process. Small gametes would have fewer nutrients to convey to the zygote, but they could move quickly and join with (i.e., fertilize) other gametes efficiently. Larger gametes could not compete with these small gametes in terms of speed, but they could compete with one another as attractive destinations for the small, fast gametes by investing even more in nutrients for the zygote. Thus, smaller and smaller or larger and larger gametes would be favored, and intermediate ones would not be (Figure 3.6).

If you study Figure 3.6, you will see that directional and stabilizing selection both reduce the variation in the population. Does this mean that eventually there is no potential for change, and therefore no potential for evolution? That is a good question. Remember the sources of variation that we reviewed early in this chapter—mutation, gene flow, and genetic mixing. These are always influential, to greater or lesser extents, and act as sources of continuing genetic variation.

Competition has figured prominently in our discussion of natural selection. The ability to compete for limited resources may strongly influence fitness; an organism that can acquire more of a limited resource than its neighbor is likely to leave more offspring, and if that ability is genetic, those offspring will also be successful. In fact, Charles Darwin, influenced by Thomas Malthus (see Chapter 1), envisioned his theory in the context of populations that inevitably exceeded resources, resulting in competition and a struggle for survival. (Indeed, a variety of unsavory

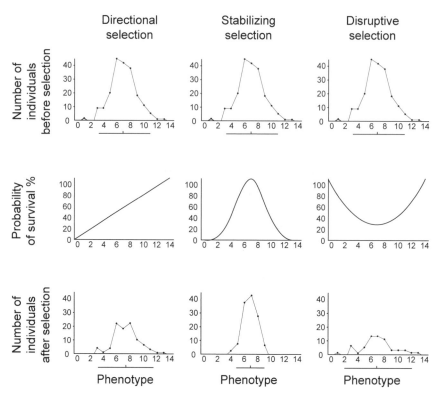

Figure 3.6 Modes of selection: directional, stabilizing, disruptive. In this figure, the "phenotype" of the population is a trait in the population that is subject to selection; in this case, the trait can be scored on a scale from 0 to 14. The top row of figures shows the frequency of different phenotypes in this population before selection. Thus, before selection, 10 of the individuals had a phenotype that could be scored as 4, 20 of them had a score of 5, etc. In all three columns, the beginning population has the same distribution of phenotypes before selection. The tic in the line below the figure represents the mean phenotype in the population; the line itself shows a measure of variation around that mean known as a standard deviation (or in this case, two such deviations). See what happens when the population is subject to *directional selection* that favors high-scoring phenotypes, as reflected in the slanted line in the second row of figures. The bottom graph shows the population after selection. Low-scoring phenotypes either disappear or are greatly diminished. High-scoring phenotypes do better. As a result, the mean phenotype shifts slightly higher. See what happens when the same population encounters *stabilizing selection*, reflected in the curved line that peaks around phenotype 7. Recall that in stabilizing selection, extreme

19th century businessmen and politicians appropriated Darwin's theory to justify their own greed, arguing that financial or political success—at any cost—was the equivalent of "fitness;" see Chapter 5.) In the case of sexual selection, organisms compete for mates and again, losers are not likely to pass on many genes.

Nice Guys Don't Have to Finish Last

If you find this portrait of nature less attractive than you would like, you are not alone. Several of Darwin's contemporaries were bothered by the implications of what they viewed as an amoral and brutal world. Alfred Lord Tennyson, who famously penned "... Nature, red in tooth and claw. . . ." asked earlier in the same poem, "Are God and Nature then at strife, That Nature lends such evil dreams?" We might well wonder, as did Darwin, if there is any room for *altruism*—that is, behavior that benefits others at a cost to oneself—in evolutionary theory. How can altruism evolve? After all, genes that produced altruistic behavior would not be passed on as readily as genes that favored more selfish actions simply because the organism with those altruistic genes would bear the cost of benefiting others, and thus have reduced fitness.

For an example, let's look at one species—ground squirrels. Belding's Ground Squirrel is a rodent that lives in large colonies. Many social animals, including these squirrels, give alarm calls that alert others in their group to approaching predators, but these calls can also be expensive because they attract the attention of the predator to the calling individual. Such calling may therefore be seen as altruistic behavior. Paul Sherman studied these animals for many years, and found that the squirrels have two kinds of alarm calls that depend on the type of predator involved. The trill indicates that a mammalian predator is nearby, and the whistle informs the colony of danger from above (e.g., hawks). Sherman observed that whistles are not very costly; the whistler is more

---

Figure 3.6 (*continued*) phenotypes are not favored (i.e., are not adaptive). The resulting population has fewer low-scoring and high-scoring phenotypes, and relatively more medium-scoring phenotypes. The mean phenotype does not shift much, but variation in the population is greatly reduced. In *disruptive selection*, the extreme phenotypes are favored, as shown by the curved line (second row) that dips in the vicinity of phenotype 7. The resulting population has far fewer medium phenotype individuals than it did before disruptive selection, and it has greater variation, even though the mean phenotype did not change much. (*Jeff Dixon*)

likely to escape the avian predator than other members of the colony are. However, the trill is expensive. An animal that trills is twice as likely to be eaten by the predator than other ground squirrels are. Trills are altruistic behaviors, and animals that trill are more likely to die than other squirrels. Natural selection should eliminate such behavior. How did trilling evolve in ground squirrels?

Darwin acknowledged that altruism was a problem for his theory, and had some ideas that addressed it. However, William Hamilton (1936–2000), in the 1960s, was the first person to propose a detailed solution. Hamilton focused on a central concept in evolution—that descendants of an ancestor will have some genes in common, passed down from that ancestor. The probability of sharing an ancestor's gene with a relative is called the coefficient of relatedness, or r. This idea is simpler than it sounds. For instance, you receive half of your genes from your mother, and half from your father—this means that your coefficient of relatedness to either of your parents is $1/2$. You have four grandparents, each of whom has given you an equal amount of genetic material, so you share one-quarter of your genes with any one of your grandparents; in that case, $r = 1/4$. On the average, you share half of your genes with a sibling, so $r = \sim 1/2$. One eighth of your genes are shared with a cousin, so $r = \sim 1/8$. You can use a pedigree to calculate these coefficients (see Calculating Coefficients of Relatedness). Given such relationships, a gene for altruistic behavior will spread—that is, be favored by natural selection—when

$$Br - C > 0 \quad \text{or} \quad rB > C$$

where B is the benefit to the recipient of the altruistic behavior, and C is the cost to the altruist, as measured by surviving offspring. This equation is called *Hamilton's Rule*, and simply says that for altruism to be favored by natural selection, the cost of the altruism must not be greater than the benefit to the recipient multiplied by the coefficient of relatedness. In other words, natural selection is more likely to favor costly altruism if it is directed at close relatives than at distant ones.

To appreciate this, let's say that your altruistic behavior toward your sister costs you one offspring, but helps her produce three additional ones. These costs and benefits do not violate Hamilton's Rule, so the behavior should be favored by natural selection and therefore can spread in the population.

$$(3 \times 1/2) - 1 > 0, \quad \text{or} \quad 0.5 > 0$$

On the other hand, the same behavior directed at a distant relative would not be so favored.

$$(3 \times 1/16) - 1 < 0, \quad \text{or} \quad -13/16 < 0$$

William Hamilton called the additional fitness that results from altruism toward a relative *indirect fitness*, in contrast to *direct fitness*, which is the fitness of an individual organism itself. *Inclusive fitness* is the combined result of indirect and direct fitness—in other words, total fitness—and natural selection that favors inclusive fitness is called *kin selection*.

This insight on the part of Hamilton explained a great many instances of altruism that we see among organisms, and that brings us back to the ground squirrels. While Sherman was studying the behavior of the squirrels, he also determined their relationships to one another. He found that the female squirrels were much more likely to trill than male squirrels were, especially if there were close relatives nearby. There is one other thing you need to know about these squirrels: males disperse during breeding season and seek other colonies. Females that are born in a colony remain there. Thus, females in a colony are related to each other, sometimes closely; males are not. The altruist ground squirrels are females, and by trilling, they benefit their relatives. Males, who are unrelated to other members of the colony, are not altruistic, and are not likely to trill. Here, yet again, Darwin's theory accounts for something in nature that is otherwise difficult to explain.

CALCULATING COEFFCIENTS OF RELATEDNESS

In the study of altruism, we are interested in the "actor" (the individual behaving altruistically) and the "recipient" (the individual receiving the benefit). We trace the paths by which genes can be shared using the pedigree: (a) In the case of full-siblings, the altruist gets half of its genes from its mother, and the same is true of the full-sibling that is the recipient of the altruism. The probability that both the altruist and recipient get the same gene from their mother is $\frac{1}{2} \times \frac{1}{2} = \frac{1}{4}$—this is because the probability that two independent events will occur is the product of the probabilities associated with each of those events (in this case, half for the actor, and half for the recipient). However, these full-siblings also share a father, and the probability that both altruist and recipient got the same gene from their father is also $\frac{1}{2} \times \frac{1}{2} = \frac{1}{4}$. Thus, there are two probabilities of $\frac{1}{4}$ each, or a total probability of half (i.e., $\frac{1}{4} + \frac{1}{4}$) that siblings share any given gene; (b) You can use the same reasoning to work out the relationships between half-siblings. In this case, there is only one shared parent (the mother or the father), and the probability that both altruist and recipient got the same gene from that parent is $\frac{1}{2} \times \frac{1}{2} = \frac{1}{4}$; (c) Relatedness between cousins is calculated in like fashion. The probability that an altruist and its parent share a gene is half, the probability that the recipient and its parent share a gene is half, and the two parents, being siblings, share approximately half of their genes. Therefore, the cousins share $\frac{1}{2} \times \frac{1}{2} \times \frac{1}{2} = \frac{1}{8}$ of their genes.

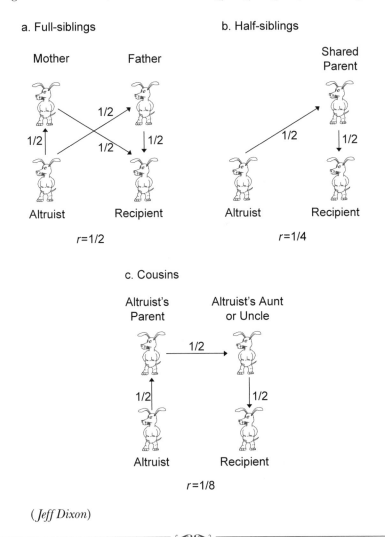

(Jeff Dixon)

Kin selection—a form of natural selection—has proven to be a powerful tool with which to understand social behavior. Kin-selected altruism, of course, occurs within a species but beneficial relationships do not end with kin selection. Mutualism occurs between individuals of two different species and benefits both. Mutualisms are abundant in nature (see Mutualism).

MUTUALISM

Mutualism abounds in nature, and includes such unlikely pairs of species as corals and algae, or flowering plants and bats. The great coral reefs would not exist were it not for tiny algae that live within the coral tissues. Corals capture food and digest it, turning it into carbon dioxide and some waste products. The algae use these compounds to make sugars, which then benefit both organisms. Much like insects, whose visits to flowers nourish the insects and disperse the plants' pollen, some bats visit flowers and gather nectar; these visitors then spread the flower's pollen to other flowers that they visit, assisting in plant reproduction in exchange for nectar. In yet other examples, ants live in associations with a variety of organisms, ranging from plants to other insects, in which the ants provide protection in exchange for nutrition. There are many things we do not understand about mutualism—for instance, how "cheating" is held in check—but it is clear that organisms frequently find ways to increase shared benefits by working together.

Remember that when we speak of attributes that are subject to natural selection and that can evolve, we mean attributes that have a genetic basis. In the case of many attributes, the environment can also play a part. Altruism is one of those traits that can have a large environmental component; we know, for instance, that animals can "do favors" for one another—that is, practice *reciprocal altruism*—and benefit from such behavior as long as they are careful of potential cheaters. And many organisms *are* careful. For example, well-fed vampire bats will give some of their food to hungry neighbors, but there's a catch: the bat that gives food to its hungry neighbor expects the favor to be returned when it is hungry. If the favor is not returned, the bat won't again share its food with the freeloading stranger.

Reciprocal altruism may have been one of the environmental influences that favored the evolution of the big human brain. In animals such as humans, with long memories, significant learning ability, and complex cultural connections, reciprocal altruism can take many delightful forms. In other words, feel free to be an altruist—in any sense of the word, not just a kin-selected one—if it makes your world a better place.

SUMMARY

Genes transfer information from parents to offspring. This information is not necessarily uniform across a population. Individuals do not look alike, and they may have different physiologies—in short, not every individual has the same genetic information. This genetic variation comes from mutation, gene flow, and genetic mixing. In addition, the relative frequencies of genes in a population do not remain constant throughout generations. Those frequencies change. Evolution is, fundamentally, a change in gene frequencies. Mutation, migration, genetic drift, and natural selection can result in changed gene frequencies. Of these four processes, only natural selection produces changes that are adaptive. Natural selection does not necessarily produce disagreeable loners. Many aspects of social behavior, including altruism, can be favored by natural selection.

4

THE SCALE AND
PRODUCTS OF EVOLUTION

At this point, you know how evolution works, but you may wonder how such a simple process could produce the living (and extinct) world. For that matter, you may wonder what that world really contains, beyond the pets and birds and flowers and vegetables we take for granted. This chapter will begin to answer these questions. We say "begin" because every day we learn more about the details of this process—more species, more relationships, and more evolutionary events. Sometimes what we learn leads us to reevaluate how closely related two species might be, for instance, or the extent of certain selective pressures. Despite such rearrangements and new ideas, the basic principles hold steady. We will present these principles in this chapter; together with natural selection, they explain the living world and how it came to be.

Evolution can be viewed at two scales—"micro" and "macro," that is, small and large, respectively. Microevolution addresses genetic change that occurs within a population or species, whereas macroevolution describes large-scale events such as the formation of species and evolutionary trends. In a way, this is an artificial distinction, because there is much overlap between the two areas. For instance, much macroevolution may result from the gradual accumulation of microevolutionary changes. Nevertheless, the distinction between micro- and macroevolution aids discussion a great deal.

MICROEVOLUTION OCCURS WITHIN POPULATIONS

Let's discuss microevolution first. To do that, we must first define "population" and "species." A *population* is a group of organisms that can interbreed; that is, the group shares a gene pool. Thus, the change in the wafflesmoozer population that we described in the previous chapter is an example of microevolution. There are changes in the frequencies of genes that regulate ear size, tail curling, whisker length, and fur color. Most microevolutionary change results from processes that we described in the previous chapter, such as natural selection, gene flow, genetic drift, and mutation; all of these processes alter gene frequencies. Nonrandom mating also changes gene frequencies. As a result of sexual selection, for instance, genes that improve an individual's chances of successful mating will be favored. Such genes are likely to become more abundant in the population than less beneficial genes.

There are many examples of microevolution, and some happen right under our noses. Take, for instance, the soapberry bug of Florida (Figure 4.1). These bugs have long mouthparts that they use to pierce the thick fruit of their native host, the balloon vine. The balloon vine has a relative in Asia, the flat-podded golden rain tree (yes, that's its name), which was imported to Florida as an ornamental plant. Soapberry bugs can eat the fruit of the flat-podded golden rain tree, and they did so, multiplying in the presence of this new resource. The covering of this new fruit is not as thick as that of the balloon vine, and after a few decades, bugs that lived on flat-podded golden rain trees had shorter mouthparts, despite being descendants of the bugs with long mouthparts that ate the fruit of the balloon vine. Of course, gardeners did not introduce the flat-podded golden rain tree to see the effect on soapberry bugs, but there was an effect, nonetheless.

We can also see microevolution occur under more purposeful conditions, as in artificial selection, for example (see Chapter 2). In the span of one human lifetime, new breeds of dogs or cattle can be developed—that is, we can produce lineages that breed true for certain genetically based traits. These lineages are developed by a simple process that has been used since the dawn of agriculture; we select organisms that have the desired trait, allow them to reproduce, and then repeat the process.

Unless you are a fan of dog shows (or rarer yet, soapberry bugs), all this may seem like so much academic fluff. You would be mistaken in that assessment, for microevolution is at work as you read this, and may even be taking food off your table. Recall that agricultural pests—mostly insects—consume over one-third of all crop plants worldwide, and are our major competitors for food. The fact that we are still engaged in that struggle against these competitors is testimony to one thing—microevolution, the evolution of pesticide resistance—which reflects genetic change in the pest population (Chapter 3). The process is simple, and ultimately Darwinian.

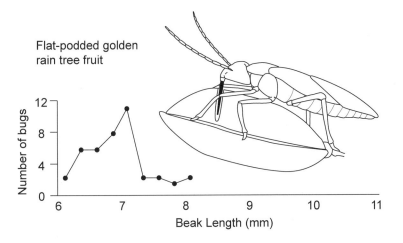

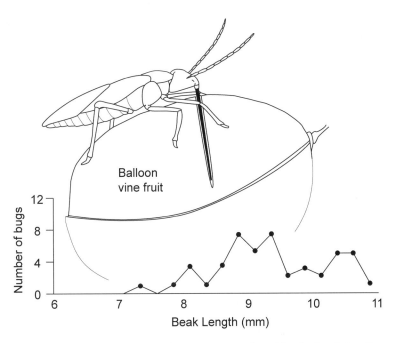

Figure 4.1 Soapberry bugs that consume the thin fruit of the flat-podded golden rain tree have shorter mouthparts than soapberry bugs that eat balloon vine fruit. Studies of museum specimens reveal that these shorter mouthparts became more frequent in the soapberry bug population after the introduction of the flat-podded golden rain tree. This is what we expect if the shorter mouthparts are an evolutionary response to a different resource. (*Jeff Dixon*)

Recall what you have learned about genetic variation (Chapter 3). If a population of insects is genetically variable regarding the ability to survive pesticides, some of them may survive pesticide application. Those that do survive will transmit those genes and that trait to their offspring. Repeated application of pesticide kills many insects at first—the dead insects are those that are susceptible to the poison. The few survivors—that is, those that happen to possess genes that enable them to survive the poison—pass this trait on to their offspring, who also survive the next application of the insecticide (and find even more food at their disposal).

The first case of insecticide resistance was reported in 1914. Less than 100 years later, there are over 500 species of insects and mites that are, as a group, resistant to a wide range of pesticides and other biologically active substances, as well as almost every killing agent that we've used. The nightmare doesn't end with those 500+ species. Fewer than 50 resistant species are beneficial. The evolution of pesticide resistance occurs much more readily in actual pest species, perhaps because predatory and parasitic insects are frequently faced by a crash in food availability when the compounds are first applied. (See Pesticide Resistance, Chapter 3.)

We see from this example that evolution is much, much more than academic fluff. Failure to take evolution into account as we design and use pesticides (and antibiotics—see Chapter 2) has created complex, expensive, and sometimes, dangerous problems.

All in all, then, when we speak of microevolution, we are speaking of changes in gene frequencies. These changes result from mutation, which is then amplified by gene flow, drift, and natural selection. Some of these genetic differences are expressed in color, size, and even behavior, and are obvious to us; others are more subtle. But how can such small details produce the seemingly unending diversity that surrounds us?

MACROEVOLUTION PRODUCES SPECIES DIVERSITY

When you ask that question, you are at the point where microevolution shades into macroevolution. Whereas microevolution addresses events that occur within species and populations, macroevolution describes events above the level of the species—that is, the relationships among groups of species and the even larger groups in which they are embedded. In addition to mutation, gene flow, drift, and natural selection (Chapter 3), macroevolution can occur as the result of large forces such as extinction (e.g., the dinosaurs).

WHAT IS A SPECIES?

Given all these possibilities, how do species form? To answer that question, we have to know what a species is. That is not as simple as it sounds. First of all, we know that members of a species resemble each other much more than any of them resembles a member of another species. For instance, bobcats are not all alike, but every bobcat is much more bobcat-like than it is lion-like. Bobcats and lions aside, we also know that many species are so similar that they are almost impossible for the casual observer to tell apart. But appearances alone do not define species. We frequently consider a species to be a group of individuals that can interbreed in nature and produce healthy, normal offspring. Indeed, if we are talking about horses or cats, that's a reasonable definition—so reasonable that it has a name: the *biological species concept*. But what about organisms that reproduce asexually, and therefore do not have to mate to produce offspring? To make things even more complicated, there are animals and plants that mate with members of the same species most of the time, but occasionally mate with closely related species, producing offspring that are called "hybrids." Are those separate species?

Scientists continue to refine the definition of a species. In the meantime, the biological species concept is not the only species concept, but it works for many organisms. The reason things are so uncertain around the concept of a species can probably be traced to the fact that a species concept is a human invention that helps us make sense of the natural world. That world has gone on for longer than we can imagine, and will continue, without much concern about our definition of a species.

HOW DO SPECIES FORM?

Species Form by Geographic Isolation, A Disturbance in Gene Flow

Let's follow our biological species concept and discover how one species can become two—that is, how species form by the process called *speciation*. Back in prehistory, the mighty Klinfizzle River changed its course, cutting directly across the range of the Greater Wafflesmoozer. As luck would have it, wafflesmoozers cannot swim very well. In fact, with the notable exception of the Wave-Riding Blue-Crested Wafflesmoozer, they sink like stones. Thus, two populations of wafflesmoozers were formed when the river changed course. It became impossible for wafflesmoozers on one side of the river to mate with wafflesmoozers on the other side. In addition, environmental conditions—including soils, vegetation, competitors, and predators—in the two areas differed. As a result, natural selection favored different traits in the two populations of wafflesmoozers. They ate different food, hid from different predators, endured different temperature challenges, and perhaps even developed different courtship displays. Over time, so many differences accumulated that even if the two groups were to reunite, what are now the Greater Crested Wafflesmoozer and the Greater Hairless Wafflesmoozer would not be able to mate successfully (Figure 4.2).

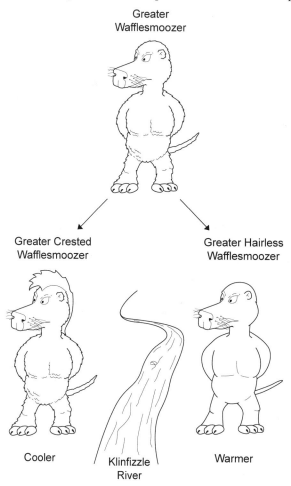

This is an example of *allopatric speciation* (*allo* = different, *patri* = land), a fancy name for a process that occurs when populations are geographically isolated. Because the world is not a uniform place, geographically isolated organisms are often subject to different selective pressures. Over time, this can result in the two populations having different traits—so different that mating would not occur between the two groups. At this point, the two groups are two species. Indeed, if founding populations are small, drift might also result in speciation.

The key here is that individuals from one of the populations do not mate with those from the other, so gene flow is eliminated. This is a situation known as *reproductive isolation*. Gene frequencies in each population change over time in

Figure 4.2 Long ago, the Klinfizzle River changed its course and split the population of Greater Wafflesmoozers, resulting in virtually no gene flow between the animals on one side of the river and the other. Given the environmental differences on both sides of the river (temperature is emphasized here), natural selection favored different traits on each side. This eventually produced two species of wafflesmoozers, each adapted to its side of the river. (*Jeff Dixon*)

response to natural selection or drift, and, because these are separate populations living under different conditions, the frequencies do not change in the same ways. As a result, the organisms in those populations differ enough so that even if they were reunited, they would not be able to mate. Complete isolation is not always essential for allopatric speciation to occur. Gene flow must be greatly reduced, but a little "leakage" does not necessarily thwart the process.

As an example, we usually think of the Isthmus of Panama as something that joins North and South America. When it was formed, however, it divided populations of marine organisms in that area. A study of snapping shrimp showed that the closest relatives of several Atlantic snapping shrimp live on the Pacific side of the Isthmus. The similar appearances of the close relatives echoed their genetic similarity. How do we know that these are indeed separate species? Scientists used mating experiments to test the hypothesis that the Pacific–Atlantic shrimp pairs were distinct species. Shrimp that did mate with related species from the other side of the Isthmus (and many were not all that interested) almost never produced offspring, thus confirming that although they were closely related, they were, in fact, separate species.

Geological changes are not the only way that a population can be subdivided and that the resulting bits and pieces can be geographically isolated. Geographic isolation can also occur when a subset of a population moves away and colonizes a different place. There is good evidence for this among Hawaiian flies, which are incredibly diverse and probably number more than 800 species. Genetic studies of these flies have shown that the most recently formed islands of Hawaii contain the most recent species.

Species Form When Shifts in Resources Disturb Gene Flow

There are also examples of potential speciation that occurs in the absence of geographic isolation. One classic story comes from flies that mate and reproduce on hawthorn plants. Almost 150 years ago, some of these flies began to use apple trees as hosts instead of hawthorn plants. Not only are the descendants of these flies genetically different from their hawthorn-loving cousins, but some of these genetic differences mean that the two sets of flies mate at different times. A different host plant, a different mating time—that is, small differences, not just rivers and volcanoes—can result in reproductive isolation. Once reproductive isolation occurs, natural selection and drift can continue to enhance the process of divergence.

Species Form When Mate Choice Restricts Gene Flow

Sexual selection can greatly influence gene flow. Recall what you learned about sexual selection in the previous chapter. What if some females in a population began choosing mates based on a different set of traits than those usually preferred by females in that population? Let's say that female Greater Crested Wafflesmoozers generally prefer males with purple crests, so that males with silver crests are less likely to find mates. If some female Greater Crested Wafflesmoozers prefer silver-crested males, and if they pass this tendency on to their female offspring, then it is not hard to imagine a group of Greater Crested Wafflesmoozers in which there are two distinct gene pools—one in which the males have purple crests, and one in which they have silver crests (Figure 4.3). If female preference in these gene pools sorted along these lines, then there would be little gene flow between the two groups. Research on a variety of insects, fish, and birds shows that sexually selected traits and the way that they differ within a population can be linked to reduced gene flow within that population.

Sexual selection can play another role in speciation. Suppose that two populations have been isolated from one another for quite some time, and that these populations are diverging, but have not yet become separate species. What happens if the two populations are somehow re-united? Some individuals may attempt to mate with individuals of the other population. If they succeed in doing so, they will produce off-spring known as *hybrids*. If the hybrids are as healthy and fertile as the offspring from within-population matings, then over time, the differences that were beginning to define the two populations may disappear. If the hybrids are less fit, however, organisms that choose mates from within their own populations and that avoid producing hybrids will be favored by natural selection. In this way, sexual selection and careful mate-choice can reduce the production of hybrids and reinforce speciation.

Species Form When Genetic Changes Disturb Gene Flow

Finally, recall the process of meiosis from the previous chapter and consider how chromosomes must line up to form new gametes. If events change the number of chromosomes, or even their shape, then gene flow can be interrupted, reproductively isolating such mutants from the parent stock. Approximately 4 percent of plant species may have originated with such meiotic and/or chromosomal accidents.

Figure 4.3 If female Crested Wafflesmoozers choose males on the basis of crest color, and if that preference is inherited and passed on to offspring, then eventually there will be virtually no gene flow between the Purple-crested and Silver-crested Wafflesmoozers. Such lack of gene flow is called reproductive isolation, and is the basis for the formation of a new species. (*Jeff Dixon*)

THE COMMON THEME OF SPECIATION IS REPRODUCTIVE ISOLATION

There are several models of speciation, ranging from those that involve great geographic separation to those that require nothing more than a different host plant. It is often impossible to witness the moment of reproductive isolation, and even more difficult to specify how much isolation is necessary to form a new species. However, we can make predictions about the kinds of patterns we would see if speciation occurs according to, say, allopatry, and we can test those predictions. (The snapping shrimp study discussed earlier in this chapter is just such a test.)

In addition, some biologists are working with "allopatric" populations of organisms that have fast reproduction times (e.g., fruit flies), investigating artificially induced speciation. What is clear in all cases is that some form of reproductive isolation is essential for speciation and that the resulting populations must diverge in response to processes such as genetic drift or natural selection.

A PHYLOGENY TRACES THE HISTORY OF SPECIATION

If you think back on what you've learned so far about speciation, you will realize that it occurs when an ancestral population diverges, resulting in two different groups. You can think of the new species as two tips on a branching tree. The speciation event itself occurs where the ancestor-branch splits into two new species-branches. "The tree of life" is a trite phrase, but an accurate one (Chapter 2). (On second thought, perhaps we would be more likely to encounter a "bush of life" if we follow our branches down into the thicket of ancestors and the ancestors of those ancestors.) The study of systematics is devoted to understanding how these branches are connected—that is, how all the organisms on Earth are related. Our ideas about these relationships are constantly being revised as more tools are added to the scientific toolbox. For instance, Charles Darwin looked at the appearance of organisms and their geographic distributions to discern evolutionary relationships; in the following century, DNA was discovered, and many scientists now examine genes to understand evolutionary relationships (Chapter 2).

As scientists explore these relationships, they develop hypotheses about them. Such a hypothesis is usually represented by a tree-like structure called a *phylogeny*, a term derived from two Greek words that mean "phylum" and "origin." Because a phylogeny is a hypothesis, it is subject to testing and refinement as more data are collected about the relationships among the organisms in the phylogeny (Figure 4.4).

How do we "read" a phylogeny? There are some important things to remember about phylogenies that reflect how scientists view evolution. First of all, phylogenies are indeed treelike; they are not stepping stones or some ever-ascending staircase as proposed by Aristotle (Chapter 1). Aristotle's idea, which came to be known as the Great Chain of Being (Figure 4.5; also see Chapter 1), has influenced our ideas about the world ever since. The Great Chain of Being represents the living world as a hierarchy—a ranked stack of organisms, piled from "lower" to "higher." (Needless to say, Aristotle placed humans at the very top.)

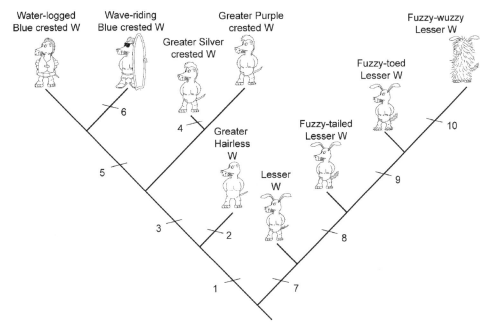

Figure 4.4 Phylogeny of wafflesmoozers. (1) Slight increase in size, reduction of ears, loss of curly tail; (2) Loss of hair; (3) Evolution of purple crest; (4) Evolution of silver crest; (5) Evolution of blue crest; (6) Evolution of aquatic adaptations; (7) Slight reduction in size, larger ears; (8) Evolution of furry tail; (9) Evolution of furry toes; (10) Entire body very hairy. W = wafflesmoozer (*Jeff Dixon*)

You can imagine that jellyfish and earthworms, along with liverworts and slime molds, would be near the bottom, while golden retrievers and chimpanzees might be near the top . . . but not as near, of course, as we are. (Medieval writers inserted the angels above us, so humans did not quite occupy the pinnacle.)

While flattering, there are some problems with this hierarchical view of life. First of all, although we might like to think that sponges, for instance, are nowhere nearly as sophisticated as we are, the truth of that smug conclusion depends on several arbitrary assumptions. It is true that sponges do not do calculus, nor have they sent a man (or even a sponge) to the moon. Indeed, they do not even have a recognizable nervous system, much less a brain. Nonetheless, they are better at being sponges than we would be at being sponges. They are certainly better at living under water, and if placed into a blender and chopped up, they can slowly reconstitute themselves, a feat that we humans would be hard

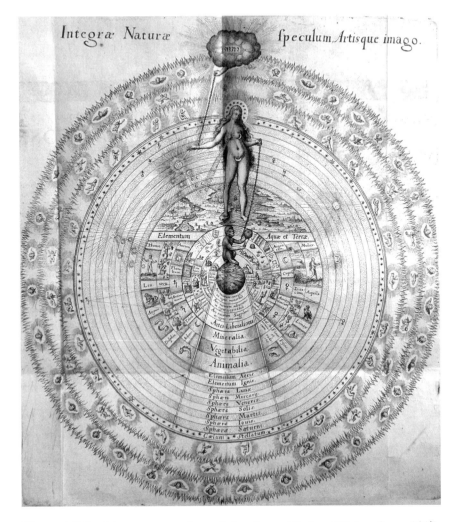

Figure 4.5 In this representation of the Great Chain of Being from 1617, species were arranged in a hierarchy according to their degree of perfection, with the most perfect organisms at the pinnacle. Each species had a clearly defined position that did not, and could not, change. (From Robert Fludd, *Utriusque Cosmi Maioris*, 1617. Image copyright History of Science Collections, University of Oklahoma Libraries)

pressed to mimic. For that matter, when the water around a so-called "water bear" (a small invertebrate called a tardigrade) dries up, the tiny aquatic animal simply gets smaller, losing internal water and producing a protective coat (Figure 4.6). Its energy requirements drop to near zero. Life is undetectable in such animals until the next rainfall, when they

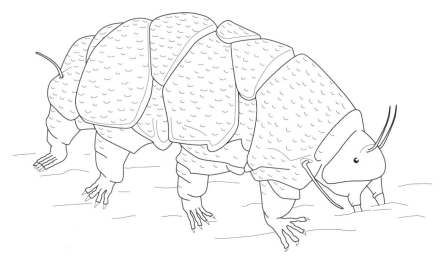

Figure 4.6 Tardigrades are tiny animals that are relatives of insects and crustaceans; most of them are no longer than a pinhead is wide. Tardigrades are well adapted to life in small amounts of water subject to evaporation (e.g., the film of water found on mosses). Tardigrades can tolerate a great deal of water loss, after which they enter a state of cryptobiosis. They pull in their legs, secrete a covering, and shut down their metabolism. Some individual tardigrades are known to have lived over 120 years in such a state. (*Jeff Dixon*)

recover and go about the business of being water bears. If we humans shriveled up and essentially became freeze-dried, we would not recover.

If you examine a phylogeny of animals (Figure 4.7), you might see that the branch leading to jellyfish (Cnidaria) splits off closer to the base than the branch leading to humans (Chordata). Indeed, things that are recognizably jellyfish-like appear earlier in the fossil record than do humans and our close relatives. If you were interpreting this according to an Aristotelian staircase model, you might infer that jellyfish were somewhere in our ancestry—say, step number 14—while we humans are at step number 72. A phylogeny is a tree, however, and not a staircase; this branching sequence merely tells us that jellyfish and humans shared a common ancestor at one time—until the time that their respective branches went their separate ways. To put it differently, *anything* alive today has been subject to evolution by natural selection just as long as *everything* alive today has been subject to evolution by natural selection. Jellyfish, gorillas, and ferns are not "older" than humans, nor are they more "primitive," and they are certainly not

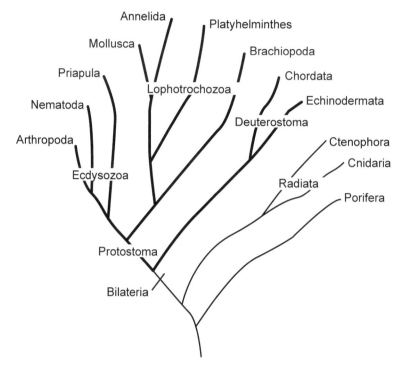

Figure 4.7 A phylogenic hypothesis showing the relationships of the major groups of animals. Such hypotheses are strongly influenced by the data that are included in (or excluded from) the analysis that produces the phylogeny. This particular phylogeny is based on some molecular data, and therefore differs somewhat from one that is based on physical traits. One challenge facing modern evolutionary biology is how to determine which data are most dependable when constructing phylogenies. Just as the wings of butterflies cannot be considered to be homologous with the wings of bats, other recently discovered traits may also require cautious interpretation until we learn more. This is typical of the scientific enterprise. (*Jeff Dixon*)

our ancestors. To greater or lesser extents, jellyfish, gorillas, ferns, and humans all share common ancestors. The branching points in a phylogeny tell us how long, relatively speaking, that shared ancestry lasted.

Aristotle was a clever, perceptive man, but there is no Great Chain of Being. Ages of evolution by natural selection mean that every surviving species is very good at exploiting its particular niche. An earthworm is no "lower" than we are, except, perhaps, in a literal sense. It is about as good at being an earthworm as anything can possibly be.

SHARED TRAITS REVEAL SHARED ANCESTRY

Organisms that share ancestry also share characteristics that are inherited from ancestors. Scientists call these characteristics *characters*. Characters can be any inherited traits, from physical appearance to a genetic code. The more characters that organisms share as a result of descent from a common ancestor, the more closely related they are. Thus, you could note that members of the group we call vertebrates have a protective covering over their spinal cords and a cranium that protects the brain. Several groups share those traits—fish, amphibians, reptiles, and mammals, for example. If all we know about these animals is that they have spinal cords and crania, we know they are related, but we do not know which animals are more closely or more distantly related. How can we find out more about the evolutionary relationships within vertebrates?

To sort out evolutionary relationships, we need to find characters that are shared by some, but not all, of the members of the *clade*. A clade (from a Greek word meaning "branch") contains all the descendants of a common ancestor. What you may notice from a study of Figure 4.8 is that with the exception of fish, all vertebrates have four limbs; that is, they are *tetrapods* (the group you met in Chapter 2). Even some snakes have vestigial limbs, evidence of their tetrapod ancestors.

As you can see, all tetrapods are vertebrates, but not all vertebrates are tetrapods. While tetrapods share the basic characters of all vertebrates, they also share characters that are specific to tetrapods. These

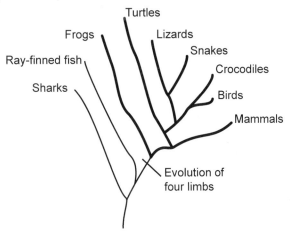

Figure 4.8 A phylogenic hypothesis of relationships among major groups of vertebrates, showing the evolution of tetrapod (four-footed) organisms. (*Jeff Dixon*)

characters are not only shared, but they are also *derived*; in other words, they are relatively new traits within the vertebrates that do not appear in the ancestor of vertebrates.

To develop phylogenetic hypotheses, scientists pay special attention to shared derived characters. These characters tell us much about evolutionary relationships. Based on shared derived characters such as four limbs, the fate of gill arches, the pelvic girdle, and many more, we now consider tetrapods as a clade within the vertebrates. To continue in this vein, some (but not all) tetrapods produce eggs (often shelled) with specialized membranes that help nourish and remove waste from the developing embryo. This clade within the tetrapods is called the *amniotes*, named after one of those specialized membranes, the amnion. The amniotes include all the tetrapods except the amphibians.

We can explore one clade after another, but the important thing to remember is that shared derived characters define evolutionary relationships. These are characters that are shared because they are inherited from a common ancestor, and the more such characters that are shared by two organisms, the more closely related they are.

THREE TYPES (BRANCHES) OF LIVING ORGANISMS

At this point, you might wish to view a bush of life that represented not just vertebrates or tetrapods, but every living group. If that's what you would like to see, then you are in for a surprise. Most of the biological diversity that we think of can be categorized into groups like "plant," and "animal," with perhaps a few "fungi" and other odd things thrown in for general interest. In reality, however, those organisms make up a small part of the living world. A recent revision of the bush of life was based on genetic material from ribosomes, the cellular structures that translate genes into proteins. Every living cell contains ribosomes, so they are good candidates for such comparisons. This analysis showed that at the very root there are three great clades: the archaea, the bacteria, and the eukarya (Figure 4.9). The exact relationship of the three is still the subject of much investigation, but the existence of three clades, emerging from early life, is well supported (see Where Did Life Come From?).

ᨆᨆ

WHERE DID LIFE COME FROM?

The short answer to this question is that no one knows for sure—in fact, "answers" are fairly speculative. Biological evolution is the study of how biodiversity came, and is coming, to be; it is not the study of life's ultimate origins. However, we do know that life on Earth originated more than 3.5 billion years ago. Chemical traces of what might have been life—that is, a self-replicating system—have been discovered in rocks that are 3.85 billion years old. We are unlikely to find traces any older than that simply because of the geological changes that have occurred in the intervening billions of years—colliding continents, erupting volcanoes, and bombarding meteors. In fact, those pesky meteors were more abundant in the early years of the solar system than they are now, and even if life had evolved during such times, it would have been vaporized, along with the oceans, in some of the big collisions.

If life originated once (or perhaps more often, only to be vaporized), why is it not originating today? For all we know, some primitive, self-replicating molecules may be assembling even as you read this, but they would be almost impossible to detect. That's because existing life—fine-tuned by eons of natural selection and evolution, and remorselessly efficient—may be busy gobbling up those defenseless morsels.

⸾

What kinds of organisms do these three clades contain? There are two fundamental cell types: prokaryotes and eukaryotes (i.e., eukarya). Eukaryotes have membranes within their cells that surround "organelles"—the machinery of the cell—including the nucleus, containing the DNA. Prokaryotes have a simpler structure, and have no internal organelles. The archaea and the bacteria are prokaryotic microorganisms that differ from one another in function and structure. The archaea are especially intriguing because some of them live in environments that cannot be inhabited by most living things—for instance, in boiling water or extremely salty pools. The eukaryota includes plants, animals, algae, fungi, and the like—all branches within the eukaryota, in addition to some eukaryotic microorganisms. While the eukaryota tends to absorb our attention, containing, as it does, penguins, oak trees, portabello mushrooms, and we

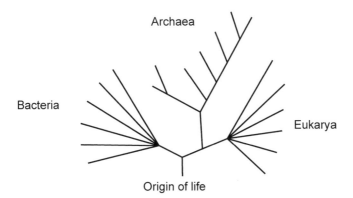

Figure 4.9 The three great clades of life. Most of the organisms with which we are familiar are Eukarya. (*Jeff Dixon*)

humans, the earth is full of prokaryotes. Not only are they the "germs" that cause disease, they help plants incorporate nitrogen and, in general, promote geochemical cycling. It was their presence on Earth for millions of years that allowed eukaryotic life to evolve.

All eukaryotic cells contain organelles called *mitochondria*, which are critical to the energy metabolism of the cell. We now know that these organelles were once a kind of bacteria called "proteobacteria" that were somehow engulfed by a distant ancestor of all eukaryotes. The proteobacterium lived, reproduced, and over evolutionary time, the eukaryotic cell and its prokaryotic guest came to rely on each other. In like fashion, an early ancestor of plants managed to engulf another bacterium, a cyanobacterium, that became the *chloroplast*, an essential organelle for turning sunlight and raw ingredients into sugar. Ancient cyanobacteria developed photosynthesis, the process by which the energy of light, together with carbon dioxide and water, is used to produce sugars and other organic compounds. Oxygen is released during this process. Accelerating photosynthesis early in the history of life probably increased atmospheric oxygen enough to allow the evolution of organisms that used oxygen to support other activities.

THE EVOLUTION OF MULTICELLULAR ORGANISMS

Although we do not wish to ignore the importance of the archaea and the bacteria, organisms that we can see tend to be those that we find most interesting, and those upon which Darwin based his observations. The timetable that you see in Figure 4.10 shows the history of these organisms.

The first multicellular animals appeared in the fossil record about 565 million years ago. Fossils from this time were first found in the Ediacaran Hills of Australia, but have now been discovered in several places. These fossils tend to be impressions of soft-bodied animals such as jellyfish and some sponges. They are fairly small, no larger than a few inches in diameter. Notice the word "diameter." These organisms, if they are symmetrical at all, are shaped like pies or cylinders. They are *radially symmetrical*, meaning they are circular, and similar all the way around, with a top and bottom (Figure 4.11). There are also trace fossils, such as burrows and the like, that indicate that at least some animals at this time may have had a head and a tail, that is, they may have been *bilaterally symmetrical*. Most animals that you know are bilaterally symmetrical; they have a head with a concentration of nervous tissue (e.g., brain and sensory organs), a tail on the opposite end, and a left and right side (Figure 4.12). Because

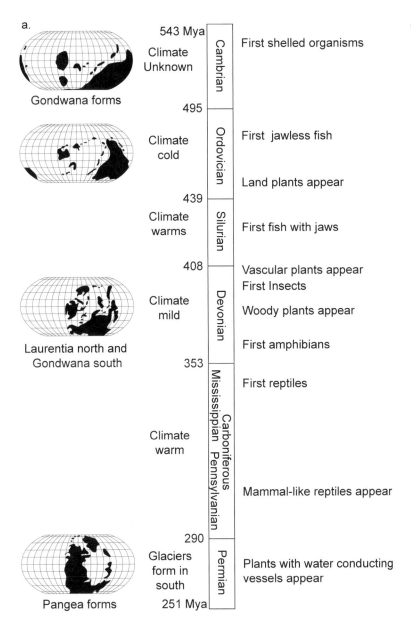

a.

		First shelled organisms
543 Mya	Cambrian	
Climate Unknown		
Gondwana forms		
495	Ordovician	First jawless fish
Climate cold		
		Land plants appear
439	Silurian	
Climate warms		First fish with jaws
408	Devonian	Vascular plants appear
		First Insects
Climate mild		Woody plants appear
Laurentia north and Gondwana south		First amphibians
353	Carboniferous Mississippian Pennsylvanian	First reptiles
Climate warm		
		Mammal-like reptiles appear
290	Permian	
Glaciers form in south		Plants with water conducting vessels appear
Pangea forms 251 Mya		

Figure 4.10 —The Paleozoic (ancient life), Mesozoic (middle life), and Cenozoic (recent life) eras are shown on this timeline. (a) The Paleozoic was characterized by a vastly different world map than the one we know today. The Paleozoic saw the birth of supercontinents Gondwana, Laurentia, and Pangea, and the evolution of most life forms. It ended with the Permian extinction. (b) In the Mesozoic, reptiles flourished and then many faced extinction. (c.) In the Cenozoic, mammals have radiated. The break-up of Pangea, which started in the Mesozoic, continued and produced the continents we know today. Mya = millions of years ago (*Jeff Dixon*)

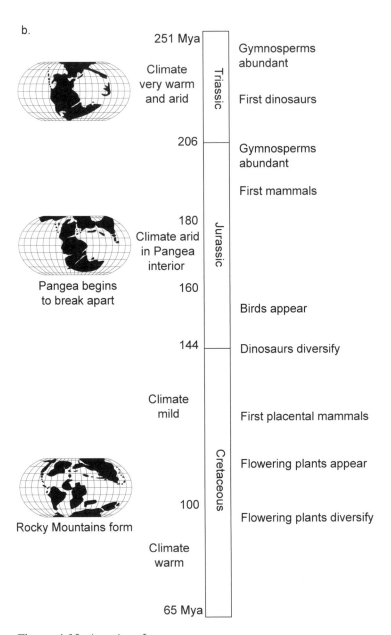

Figure 4.10 (*continued*)

c.

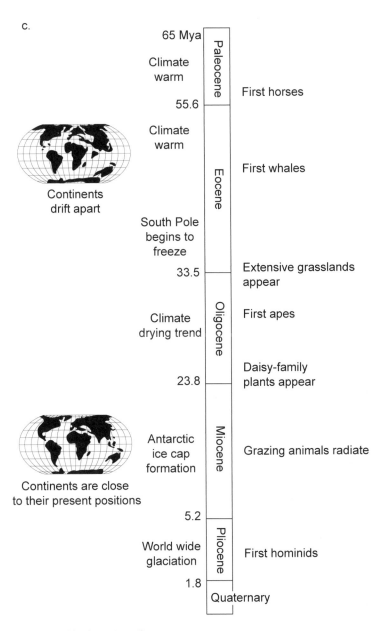

65 Mya	Paleocene
Climate warm	
55.6	First horses
Climate warm	Eocene
	First whales
Continents drift apart	
South Pole begins to freeze	
33.5	Extensive grasslands appear
Climate drying trend	Oligocene
	First apes
	Daisy-family plants appear
23.8	
Antarctic ice cap formation	Miocene
	Grazing animals radiate
Continents are close to their present positions	
5.2	
World wide glaciation	Pliocene
	First hominids
1.8	
	Quaternary

Figure 4.10 (*continued*)

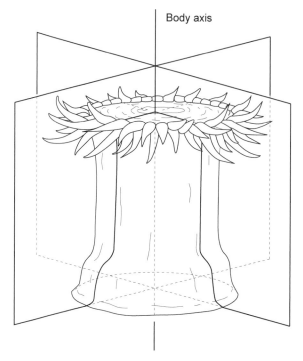

Body axis

Figure 4.11 Radial symmetry is circular, like the symmetry of pies and soft-drink cans. No matter how you slice something that is radially symmetrical, you will get two identical halves, as long as the slice runs in the same direction as the axis of the body. (*Jeff Dixon*)

these animals left only trace fossils, we conclude that they had no shells or other hard parts with which to leave more substantial remains. In addition, the burrows that they did leave are fairly simple, suggesting that they may have faced few ecological challenges.

It is the slightly younger (in geological terms) fossils that are truly amazing. From 543 to 506 million years ago, an incredibly diverse group of animals suddenly appeared on the scene. These animals were bilaterally symmetrical; they were large, relative to the Ediacaran fossils, and among them we see evidence of segmentation, limbs, shells, exoskeletons (the crunchy covering of insects and crustaceans), and even notochords, the flexible rod-like signature of our own phylum (Figure 4.13). They are the early members of many existing phyla, exhibiting many traits that are present in today's molluscs, crustaceans, worms, and primitive fish (see Appendix 5: The Products of Evolution). These fossils were first discovered in the early part of the 20th century in British Columbia, Canada, in a formation called the Burgess Shale. A similar deposit has been found in China. The fossils are from a period of time over 500 million years ago called the Cambrian, and their abundance and diversity have amazed scientists, causing this expansion of life forms to be known as the Cambrian Explosion. Most of the body plans of modern animals appeared during the Cambrian.

Why such an "explosion"? The Earth was changing, as it always has. Growing populations of photosynthetic algae meant that the atmosphere contained more oxygen. With more abundant oxygen, animals that used oxygen to fuel locomotion were able to be more active and

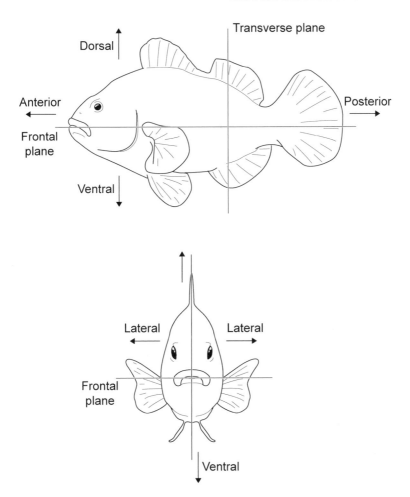

Figure 4.12 Bilateral symmetry works well for mobile animals because it can produce a streamlined form, with nervous tissues concentrated in the anterior end. There is only one way to slice a bilaterally symmetrical object that will yield two identical halves. (*Jeff Dixon*)

support more tissues and more complexity. The animals that left impressions in the Ediacaran Hills were either sedentary or drifted through the ocean, filtering or trapping small food particles. In contrast, the animals of the Cambrian were on the move, seeking prey, fleeing predators, and developing defenses that left marvelous fossils, some of which still bear the marks of predator attacks. [By the way, plants have experienced "explosions" of their own. The two most conspicuous ones involved the

Figure 4.13 Burgess shale fossils are highly varied, and give us a glimpse of the "explosion" of life that occurred during the Cambrian. (*Jeff Dixon*)

invasion of land from a previously aquatic environment (about 400 million years ago), and the evolution of flowers, a major step in reproduction that occurred about 110 million years ago.]

All of the major body plans of animals can be found in the Cambrian Explosion (for more about body plans, see Appendix 5: The Products of Evolution). Insofar as we know, the actual species of the Cambrian are extinct, but many of their descendants are with us, and can be recognized by the existing modifications of those Cambrian body plans. Although there are countless living species, this abundance actually represents modifications of only a few highly successful body plans. The presence/absence and type of body cavity, appendages, body covering—these and other basic traits have been refined by natural selection and have proliferated through adaptive radiation (see below). The animals that bear these traits can be found in almost every environment, from the deepest hydrothermal vents of the ocean to the ice of montane glaciers.

ADAPTIVE RADIATION FUELS BIODIVERSITY

The Cambrian Explosion happened in a past that we are just beginning to understand, and the speed with which it happened is still being debated. We do know, however, that biological change can occur rapidly, and in response to ecological opportunity. The so-called Darwin's finches of the Galápagos Islands give us one fairly modern example of such *adaptive radiation*. You met these birds in Chapter 2.

Adaptive radiation happens when a species gives rise to several descendant species. This process often occurs when the ancestral species encounters a novel environment in which there are *niches* that have not been occupied. (In ecology, a "niche" refers not only to the place that an organism lives, but how it makes its living. As ecologist Eugene Odum pointed out, if the habitat is an organism's address, then the niche is its profession.) As with Darwin's finches, different niches favor different adaptations.

The ancestors of Darwin's finches were ground-dwelling, seed-eating birds that found their way to the Galápagos Islands about 3 million years ago. As with any group of animals, there was within-species variation. Some finches were better at eating insects than they were at eating seeds, and others were better at eating smaller seeds than larger seeds, and some survived better in the trees, others in cactus. Over time, the ancestral species gave rise to at least 13 descendant species (Figure 4.14).

There are many examples of adaptive radiation, and not all of them depend on an ancestral species finding its way to a distant island. Sometimes, evolutionary novelties can be remarkably successful. Nowhere is this as apparent as with the great arthropod clade. Arthropods are animals with exoskeletons, and include insects, crustaceans, mites, spiders, centipedes, and other animals too numerous to mention. Most animals on Earth (over 75 percent) are arthropods, and the evolution of that crunchy exoskeleton, along with an incredible variety of crunchy appendages worthy of a Swiss Army knife, is what made that diversity possible. Arthropods use those appendages to grab, crush, swim, breathe, mate, protect young, walk, run, sting, spin silk, and do other things you can't imagine. Arthropods did not need the Galápagos Islands to

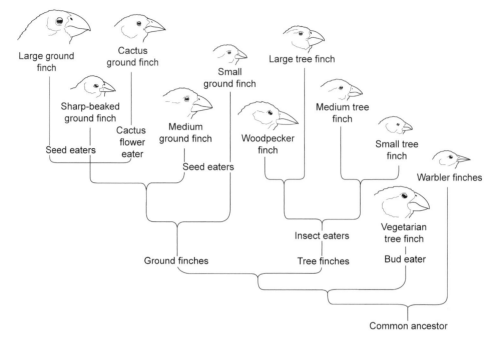

Figure 4.14 This is a partial phylogeny of the Galápagos finches. South American finches colonized the islands, and over time, their descendants underwent adaptive radiation, exploiting a variety of habitats and food sources. You can see this radiation represented in their size differences and the shapes of their beaks. (*Jeff Dixon*)

undergo adaptive radiation. They simply needed an exoskeleton, some appendages, and genetic variation.

Adaptive radiation also occurs when niches become empty because their previous occupants have disappeared. The species of the Burgess Shale are no longer with us. Neither are the massive dinosaurs, nor, for that matter, the tiny animals that gave rise to mammals. Extinction happens, and when it does, niches are opened for adaptive radiation to fill with new species. Sometimes, however, extinction happens with a vengeance.

POISONOUS OCEANS, DEATH STARS, AND MASS EXTINCTION

On at least five different occasions during the last 510 million years, more than 60 percent of existing species were extinguished in the space of a million years (Figure 4.15). These events affect large numbers of species and occur rapidly and globally. They are called *mass extinctions*.

What causes a mass extinction? In general, these extinctions are caused by drastic changes in habitat. We are happy to report that there have not been enough mass extinctions to generate a general pattern of what causes such massive habitat change, but we do understand that habitat change is at the core of these dramatic events. We will discuss the hypotheses surrounding two of the better understood extinctions.

The largest extinction occurred 250 million years ago, at the end of the Permian. Through completely natural means, the ocean might have become toxic at this time. As you may know, the continents are not stationary, but float very slowly, a phenomenon called *continental drift*. At the end of the Permian, the continents had just come together and formed a single large continent called Pan-

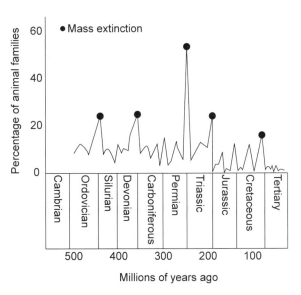

Figure 4.15 Five mass extinctions. (1) The Ordovician extinction (440 million years ago) eradicated many species of small organisms that lived on the bottom of the ocean. (2) The Devonian extinction (365 million years ago) killed small marine organisms and coral reefs. (3) The Permian extinction (250 million years ago) eliminated more than 90 percent of species on Earth. (4) The Triassic extinction (215 million years ago) wiped out more than 20 percent of existing families. (5) The Cretaceous extinction (65 million years ago) killed more than two-thirds of the existing species. (*Jeff Dixon*)

gaea. Needless to say, this affected ocean currents, which in turn, had an enormous effect on climate, just as they do today. (There is evidence that during this extinction period, glaciers formed and retreated multiple times—a sign of major climate fluctuation.) If the currents could not provide enough oxygen to ocean depths to sustain life, these deep waters would be stagnant and would lack oxygen. If surface photosynthesis persisted, then atmospheric carbon dioxide would be depleted, and global temperature would decrease. Advancing glaciers would increase ocean circulation, which in turn would bring the deep ocean waters to the surface, poisoning surface life, releasing carbon dioxide to the atmosphere, and increasing temperature. This cycle probably continued until Pangaea fragmented. This extinction had a dramatic impact on marine animals; 90 percent of marine invertebrate species disappeared, supporting the notion of a toxic ocean.

In contrast, the cause of the late Cretaceous extinction came from outer space. Sixty-five million years ago, a meteorite 10 kilometers in diameter struck the earth in northwest Yucatan, Mexico. All over the world, the noble metal iridium—an element that is rare on Earth but abundant in meteors—is found at the boundary of the Cretaceous and the Tertiary (called the K-T boundary). Based on this observation, Luis Alvarez and coworkers suggested that a meteor strike might have caused the K-T extinction. Initially, this suggestion was viewed with a great deal of doubt and even ridicule. However, evidence began to pile up. For instance, tiny grains of melted and recrystallized minerals are found all over the Caribbean. They appear to have been melted by the heat of the meteor and ejected. From their characteristic shape, we know that they cooled as they splattered away from the impact site. At last, in the early 1990s, a crater 180 kilometers in diameter was discovered near Chicxulub, Yucatan, Mexico. The impact happened. What did it do to living things?

First of all, it is important to note that not all large meteors cause mass extinctions when they hit the earth. We have evidence of other impacts that are not correlated with mass extinctions (Figure 4.16). However, a meteor impact could affect life on Earth in many ways. Vaporization of compounds would alter atmospheric chemistry, and that, together with vast amounts of floating debris, would result in global cooling. There is evidence of widespread fires accompanying the impact, which would have produced soot and added to global cooling. The impact itself may have increased volcanic action, thereby producing more atmospheric disturbance. Finally, a 300-kilometer long sandstone deposit through Texas and Mexico bears witness to a massive tsunami that must have followed the impact.

The idea here is not so much that the mass extinction was caused because organisms were squashed by an extra-large meteor, but rather because the meteor disrupted climate, geochemical cycles, ocean currents, and ultimately, ecological interactions. It caused dramatic changes in habitat. These changes had long-term extinction effects that lasted for half a million years.

The subsequent recovery took five times that long—2.5 million years. Even as the Permian extinction made room for reptiles, the obliteration of most dinosaurs at the K-T extinction allowed the adaptive radiation of mammals (Figure 4.17).

Figure 4.16 The late Cretaceous mass extinction was caused by a meteor that hit Earth in what is now Mexico. However, not all meteorites cause mass extinctions. The Barringer Meteorite Crater shown here is a mile-wide, 570′ deep hole in the arid sandstone of the Arizona desert. When the crater was discovered, the plain around it was covered with chunks of meteoric iron covering an area 10 miles in diameter. Today, thousands of small meteorites weighing a quarter of a pound or more hit Earth every year.

The story of mass extinction is not over. Current extinction rates are 100–1,000 times the normal rate of extinction, even without the assistance of misplaced ocean currents or rogue meteors. If these rates continue, then they will culminate in another mass extinction. We understand current extinctions far better than those in the distant past. Today's extinctions are caused by human-mediated habitat destruction.

SUMMARY

Evolution occurs at different scales, both within populations (microevolution) and as it generates new species (macroevolution). Species can form when gene flow is interrupted by such things as physical impediments, habitat diversification, sexual selection, and genetic change. Using shared traits, which reveal shared ancestry, biologists construct phylogenies to trace the history of speciation and the relationships among organisms. Phylogenies are hypotheses about relationships;

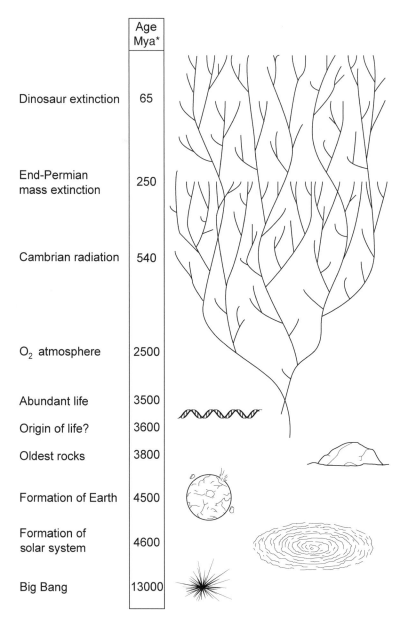

	Age Mya*
Dinosaur extinction	65
End-Permian mass extinction	250
Cambrian radiation	540
O_2 atmosphere	2500
Abundant life	3500
Origin of life?	3600
Oldest rocks	3800
Formation of Earth	4500
Formation of solar system	4600
Big Bang	13000

Figure 4.17 Extinction and radiation. See how the bush of life is pruned at times. *Mya = millions of years ago (*Jeff Dixon*)

because the traits that are used in creating a phylogeny are clearly defined, the phylogenetic hypothesis can be tested as new traits are discovered and added. The resulting "bush of life" has three main branches (i.e., the archaea, bacteria, and eukarya), and those branches continue to subdivide as a result of adaptive radiation. Extinction creates new opportunities for adaptive radiation and species formation. Of the species alive today, approximately 1.8 million have been described, and perhaps up to 200 million await description.

5

EVOLUTION AND OUR DAILY LIVES

Darwin's theory of evolution provides a foundation in biology so basic that many biologists often take it for granted. Despite being almost 150 years old, the theory of evolution by natural selection continues to help us make sense of the world and accurately predict biological discoveries. For example, as you learned in Chapter 3, species differ in their degree of sexual dimorphism, largely as a result of sexual selection. If you were to come upon a species that showed very little sexual dimorphism, you could be fairly certain that in that species, an individual is unlikely to have very many mates, at least not in the same breeding season. Similarly, if you were to come upon a species in which sexes were wildly dimorphic, a good guess would be that monogamy is not the rule. These and many other predictions emerge from Darwin's great idea (see Darwin's Predictions Continue to Ring True). Indeed, medical research has deep roots in evolutionary theory: treatments that work on other primates are much more likely to be effective on humans than treatments that alter the physiology of turtles, fish, or spiders, but do not affect primates. The fact that some animals are good models for human disease has been a boon to veterinary medicine; our pets and livestock have benefited greatly from discoveries made in the pursuit of human health.

DARWIN'S PREDICTIONS CONTINUE TO RING TRUE

There have been many dramatic confirmations of Darwin's theory of evolution by natural selection. In this book you've learned about some of them—for example, *Archaeopteryx* (Chapter 2) is a transitional form between reptiles and birds, and the moth *Xanthopan morgani praedicta* was found after Darwin had predicted its existence (see Chapter 1, Darwin's Bold

Prediction). Yet another confirmation of Darwin's theory occurred in 2006 when biologists discovered fossils of what came to be known as "fishapods."

Darwin's theory predicts that fish gave rise to four-legged animals known as tetrapods about 375 million years ago. Knowing this, a group of biologists from the University of Chicago went to Ellesmere Island in the Canadian arctic (only about 900 miles from the North Pole) to search for evidence of this evolutionary transition. Why Ellesmere Island? Remember from Chapter 2 that finding a fossil means finding a rock likely to contain the fossil. Ellesmere Island has large amounts of fossil-bearing rocks of red siltstone that date back to 375 million years ago—just about the time when the transition of animal life from water to land was suspected to have occurred. Moreover, 375 million years ago Ellesmere Island was a subtropical river delta (continental drift later moved it to its present location).

At Ellesmere Island the biologists found fossilized fishapods. Like fish, these 1.3-meter-long animals had gills, scales, fins, and a snout. However, like land animals (and unlike fish), the fishapods had flexible necks, eyes atop broad alligator-like skulls, wrists, and limb-like fins having a flexible elbow joint, a primitive wrist, and five fingerlike bones. The fishapods also had strong interlocking ribs, which suggested the presence of lungs for breathing air on land. As a result of these adaptations, fishapods could lift themselves out of the water, and thereby have a better view of potential prey and predators.

The biologists named their discovery *Tiktaalik roseae* (in the language of the local Inuit, *tiktaalik* means "large fish in stream"). This organism is exactly the sort of transitional animal predicted by Darwin's theory; it fits anatomically between lobe-finned fish (such as *Panderichthys*) and primitive tetrapods (such as *Acanthostega*). Like *Archaeopteryx* and other so-called "missing links," *Tiktaalik* supports Darwin's conviction that "endless forms most beautiful and most wonderful have been, and are being, evolved."

⎯⎯⎯⎯⎯⎯⎯⎯⎯⎯⎯⎯ ᥣᵔᴏᥣ ⎯⎯⎯⎯⎯⎯⎯⎯⎯⎯⎯⎯

The reverberations of Darwin's theory show no sign of diminishing. Advances in genetics have pinpointed the molecular changes that have driven evolution. For instance, the genetic information carried by chimpanzees—our closest relatives—is virtually identical to that of humans; biologists anticipate that comparing this information—that is, conducting a detailed study of the 40 million evolutionary changes that distinguish humans from chimpanzees—will help us determine the genetic differences that make us human. This is much more than an academic exercise. Indeed, these studies could improve our lives because humans are susceptible to coronary heart disease, viral hepatitis, malaria, AIDS, and other ailments that are usually not a problem for chimpanzees. Knowing the evolutionary reasons for these differences will help us learn why we are more susceptible to these diseases, and may lead to cures. Similar applications of evolution abound in modern society.

As you will see later in this chapter, evolutionary considerations are crucial as we deal with modern problems. That said, every great idea has its misapplications; one only has to contemplate the excesses of communism, unbridled capitalism, and some religious notions to see the sad truth of that statement. Darwin's idea has fared no better. When misunderstood or applied incorrectly (with perhaps ulterior motives), good science can produce undesirable outcomes. In the past, Darwin's ideas have been misapplied to social settings with disastrous results. Two such examples are Social Darwinism and the eugenics movement.

SOCIAL DARWINISM

The Enlightenment's thinkers who preceded Darwin often speculated that societies progress through stages of increasing development. Darwin himself argued that societies have been strengthened by social instincts such as sympathy and moral sentiments that evolved through natural selection. In Darwin's day, a group of capitalists and philosophers— most notably Herbert Spencer, who, like Darwin, was inspired by Malthus (Chapter 2)—promoted using the "survival of the fittest" principle in British businesses. When these people saw the parallels of their work and Darwin's theory, they immediately began to apply what they believed were Darwin's lessons to society. These so-called Social Darwinists wanted to integrate evolution and natural selection into human cultural systems by promoting social policies that would allow weak and unfit individuals to fail and die, while helping the strongest and fittest people to flourish. The Social Darwinists preached a relentless, harsh message—namely, that people got what they deserved, that the fittest not only survived, but prospered, and that society should eliminate unfit individuals because nature, in fact, does the same thing. (The fact that these Social Darwinists had both survived and prospered was, of course, entirely coincidental.)

Virtually all of the early Social Darwinists, including robber-barons Andrew Carnegie (the founder of U.S. Steel) and John D. Rockefeller of Standard Oil, were capitalists who believed that the rich and powerful were better adapted to the social and economic climate than the weak and poor. To these people, such a conclusion was not merely logical and natural, but morally right. They used Darwin's views to justify colonialization (of weaker, unfit indigenous peoples), militarism (the casualties must be unfit people), and racial discrimination (against allegedly unfit groups). Although Darwin rejected claims that evolution was progressive and goal-oriented, Spencer believed that Social Darwinism would lead to a final, and perfect, society.

Social Darwinism became popular late in the 19th century and continued to the end of World War II. However, Social Darwinists' reasoning faltered when they used the so-called "naturalistic fallacy" to derive *ought* from *is*. Indeed, someone who *is* afflicted with diabetes is not necessarily someone who *ought* to have diabetes. Social Darwinism had virtually nothing in common with Darwin's theory of evolution; Darwin never advocated extending natural processes into human social structures. Nevertheless, Spencer's influence was large, and much of what was ultimately attributed to Darwin can be traced to Spencer's philosophical shifts.

Social Darwinism claimed that society would regulate itself—for example, the poor might reproduce more, but they would also have a higher mortality rate. However, not everyone was patient enough for this type of regulation. Instead, many people wanted to produce better societies via governmental interventions that directed human evolution. These people founded the eugenics movement.

EUGENICS

The most extreme examples of Social Darwinism became known as eugenics (derived from the Greek word meaning "good origin" or "good breeding"). Eugenics is a social philosophy that advocates improving human genetic qualities via social interventions (e.g., birth control, selective breeding) that make people smarter, save resources, and diminish suffering. Selective breeding had been advocated as far back as Plato, who wrote in *The Republic* that "The best men must have intercourse with the best women as frequently as possible, and the opposite is true of the very inferior." (Plato also urged using a fake lottery so feelings were not hurt by the selection criteria.)

The modern discipline of eugenics was founded by Sir Francis Galton (Charles Darwin's cousin) in 1865. Galton believed that because artificial selection could be used to exaggerate traits in animals, we could expect similar results with humans, arguing that "it would be quite practicable to produce a highly gifted race of men by judicious marriages during several consecutive generations." Galton, worried that natural selection was being thwarted by misguided programs that encouraged "unfit" people to have children, advocated policies that would save society from what he termed a "reversion toward mediocrity." (Galton had coined this phrase in his studies of statistics, and the concept later became the now well-known "regression toward the mean.") Galton believed that "character, including the aptitude for work, is heritable like every other faculty." Galton summarized his attempts to make a science out of human breeding this way: "What Nature does blindly, slowly, and ruthlessly, man may do providently, quickly, and kindly."

Galton's goal of improving the human race "through better breeding" attracted many supporters (including Woodrow Wilson and Alexander Graham Bell), and became very popular in the first half of the 20th century. In 1896, Connecticut became the first state to enact marriage laws with eugenic criteria; these laws banned "epileptic, imbecile, or feeble-minded" people from marrying. Many other states followed suit, and the results were catastrophic, for between 1907 and 1963, more than 60,000 "imbeciles" were forcibly sterilized in the United States. Virtually all non-Catholic western nations—including Sweden, Canada, Australia, Norway, Finland, Denmark, and Switzerland—applied similar sterilization policies to "hereditary and incurable drunkards, sexual criminals, and lunatics." Organizations such as The Race Betterment Foundation became common, and in the United States, state fairs often gave awards to the largest and "best" Caucasian families.

Darwin opposed the tenets of Social Darwinism and eugenics; he noted in *The Descent of Man* that eugenics would be an "overwhelming present evil" and that humans should not "check our sympathy, even at the urging of hard reason, without deterioration in the noblest part of our nature." However, eugenics was supported by Darwinists such as Fisher and Haldane (Chapter 1), and at many colleges and universities (including notable Ivy League institutions), eugenics became an academic discipline. In public schools across the United States, students were often indoctrinated about eugenics in their classrooms. For example, the biology textbook used in 1925 by John Scopes (Chapter 2) discussed the "five races" of man and assured Scopes' all-white, legally segregated high school students that Caucasians are "the highest type of all." The textbook, with an avowed goal of improving the future human race, proposed eugenic remedies for societal ills. After discussing the inheritance of crime and immorality, the author proposed an analogy:

Just as certain animals or plants become parasitic on other plants or animals, these families become parasitic on society. They not only do harm to others by corrupting, stealing or spreading disease, but they are actually protected and cared for by the state out of public money. Largely for them the poorhouse and the asylum exist. They take from society but they give nothing in return. They are true parasites.

This analogy led to a "remedy":

If such people were lower animals, we would probably kill them off to prevent them from spreading. Humanity will not allow this, but we do have the remedy of separating the sexes in asylums or other places and in various ways preventing intermarriage and the possibilities of perpetuating such a low and degenerate race. Remedies of this sort have been tried successfully in Europe and are now meeting with success in this country.

These "remedies" included forced sterilizations. The U.S. Supreme Court's *Buck v. Bell* decision in 1927 endorsed such sterilization, which was practiced in 27 states.

Meanwhile, in Germany, Darwinist Ernst Haeckel's claim that "politics is applied biology" was used by Nazis to justify their militant naturalism and racism. Haeckel welcomed the suppression of "less advanced" races because he believed they were relics of earlier stages of human evolution. As he wrote in his book *History of Creation,*

We are proud of having so immensely outstripped our lower animal ancestors and derive from it the consoling assurance that in the future also, mankind as a whole will follow the glorious career of progressive development and attain a still higher degree of mental perfection.

According to Haeckel and others, helping the poor and disabled was a waste of time and a recipe for added suffering later, for these "good-for-nothings" were unfit and genetically predestined to fail. For Haeckel, the "lowest" humans were various "species" of Africans and New Guineans, and at the summit were Europeans, which he designated *Homo mediterraneus.* And within *H. mediterraneus,* Haeckel's fellow Germans were at the pinnacle. Sooner or later, most of the other races would "succumb in the struggle for existence to the superiority of the Mediterranean races."

German government leaders used eugenics and Haeckel's vision of human destiny to justify attempts to maintain a "pure" race via "racial hygiene." The Nazis forcibly sterilized hundreds of thousands of "unfit" people, and euthanized tens of thousands of institutionally disabled people. Awards were given to "Aryan" women who had large families, and "racially pure" single women were impregnated by governmental officials. The equipment used for the mass-killings at Nazi concentration camps was developed by the German eugenics program, and at the Nazi War Crimes Trial at Nuremberg (a town where more than 450,000 "defectives" were forcibly sterilized), defendants cited the United States' eugenics program as an inspiration for their atrocities. In terms of sheer numbers of suffering people, the Nazi political and social machine produced the most nightmarish mass torture and homicide in history. As you might imagine, the popularity of eugenics plummeted after the extent of their atrocities was fully exposed. Meanwhile, it is worth noting that antievolutionists count among their number the Ku Klux Klan and similar hate-groups; Klan members were horrified at the thought that people of color, or simply of different ancestry, might be relatives, however distant.

THE EVOLUTION OF PATHOGENS

While eugenics pretends to concern itself with health improvements, Darwinian theory makes real and important contributions to human health, simply because pathogens, like the rest of us, evolve. Physicians must learn mountains of information before being loosed upon their patients, so it is not surprising that medical schools have not been at the forefront of evolutionary research. That is changing, as the importance of evolution to human health becomes clear. For instance, traditionally, medical researchers assumed that parasites and pathogens evolved in a way that would minimize damage to their hosts. This reasoning was not without merit; researchers assumed that if the host lived, the parasite would be better off than if it were in a dead host. What they did not take into account was natural selection and parasite fitness.

Beginning in the 1970s, Robert May and Roy Anderson began to question those assumptions. Pathogens do not always evolve toward benign coexistence. It may be to a pathogen's advantage to cause illness or even death if that increases parasite fitness. As unlikely as it may seem at first glance, host illness can increase parasite fitness if it increases parasite transmission. Remember the last time you had a bad cold; you were coughing, you were sneezing, you were sick—and while you were clearing out your nasal passages, you were spreading cold viruses with every sneeze and cough. Anderson and May stated that there is no single outcome for pathogen evolution; rather, whether a pathogen becomes more or less virulent (that is, disease-causing) depends on multiple aspects of parasite and host biology, including things like parasite transmission and host immunity. For some parasites, a sick or dying host is the best way to get around; for others (think sexually transmitted diseases), the glow of seeming health is a great ticket to the next host. In the first case, we expect the parasite to evolve increasing virulence; in the second, we expect diminished, or delayed, virulence.

Some pathogens evolve rapidly because they undergo rapid mutation. Every time a flu vaccine is manufactured, the pharmaceutical company that produces it is making an educated guess about the evolution of flu virus. Flu virus mutates quickly, and many scientists study this virus in an attempt to predict the next strain so that the vaccine can be more accurately designed.

Meanwhile, the evolution of flu virus has taken a potentially ominous turn. In the early part of the 19th century, a type of flu emerged that killed millions of people. In the face of news about a new strain of flu from birds ("avian flu") that was often lethal in humans, scientists realized that they could look at the genetic material from the 1918 strain and see if the current avian flu was similar to the killer flu of decades past. Thus, in 2005, biologists used tissue from a body frozen in Alaska since that epidemic to sequence three genes from that 1918 flu virus. They found that the virus began as an avian virus, and that a few mutations enabled it to spread among humans. Similar strains of avian flu have appeared since 1918 on at least three occasions, with less disastrous results. Today's news stories about avian flu represent fears that the events of 1918 may repeat themselves.

Human evolution has offered a changing landscape to disease organisms, and one that increasingly welcomes them. First of all, we have domesticated a variety of animals—and probably as a result, share scores of their diseases. Our agricultural practices make us a predictable source of blood meals for mosquitoes, which transmit a variety of blood-borne pathogens; nomads are not so predictable, and in addition, the standing water generated by many agricultural practices favors mosquitoes and snails, the vector of blood flukes. As sedentary farming ways gave rise to villages, which grew to towns and cities, an increasing number of epidemic diseases were able to establish themselves. For instance, measles requires at least a few thousand new (nonimmune) hosts per year to be harbored in a population, otherwise it will go locally extinct. As cities grow, however, the probability of such local extinction at this point in human history is vanishingly small. With every shift in human social behavior, natural selection favors pathogen change, and routes that are open to those pathogens shift as well.

Emerging diseases, HIV evolution, antibiotic resistance—evolution has profound implications for human health. Even the health problems associated with aging can be seen in evolutionary terms; after all, the older a person gets, the less likely he or she is to reproduce, and whatever happens to health at that point is subject to little, if any, natural selection.

In *The Evolution Explosion*, Steve Palumbi presents the challenge this way:

Furthermore, if drug resistance is inevitable, then by choosing drugs, we are in effect choosing the evolutionary trajectory of the virus. Why not use this opportunity to channel the virus into an evolutionary cul-de-sac and then let loose the pharmaceutical dogs? Granted we do not know how to do this yet. But while HIV evolves, it thwarts us. And so we must learn how to control evolution in order to survive the evolutionary skills of HIV.

The same can be said for many pathogens, as well as for many problems that humans face. Humans are agents of massive environmental change, and are therefore—whether or not we choose to think about it—agents of evolutionary change. We might as well be smart about it.

SOCIOBIOLOGY AND EVOLUTIONARY PSYCHOLOGY

In 1975, Harvard biologist E. O. Wilson published *Sociobiology, The New Synthesis,* a monumental undertaking which examined social behavior across living animals, and how it may have been shaped by natural selection. Wilson intrepidly included human behavior in his analysis, and this generated a firestorm of criticism. Evolutionary biologists such as Stephen Jay Gould attacked Wilson's controversial argument, claiming that it was merely an endorsement of the social status quo. Today, the field of evolutionary psychology uses Darwin's theory to explain human behavior.

At this point, our understanding of the interaction of human genetics with behavior is limited. We know, for instance, that the actual expression of any gene is influenced not only by that gene, but also by the environment in which the expression takes place. One of the great challenges of genetics is to figure out what part of any given trait is inherited and what part is environmental. Nowhere is this challenge greater than in the area of human traits, and especially human behavior.

HUMAN EVOLUTION

At the heart of eugenics is the idea that humans evolve—and indeed, we have, but perhaps not in the way you imagine. Some people envision the ancestor of humans to be like an ape—perhaps like our closest relative, the chimpanzee. Remember the lessons of Chapter 4—nothing alive today is an ancestor, and everything alive today has been subject to natural selection for as long as everything else. Apes may be our distant cousins, but they are not our grandparents, many times removed.

If we look at the "bush of life," we see humans on a branch called "Primates." This branch contains monkeys, apes, lemurs, and humans, all of which share traits like forward-facing eyes, opposable thumbs, and a fancy cerebral cortex (the part of the brain that is very good at controlling voluntary behavior and complex mental tasks). We are mostly omnivores—that is, we are not restricted to meat or vegetation in our diets—a fact you may notice as you think about your next snack.

Leaving the lemurs and another group called the tarsiers aside, most primates belong to a group called the anthropoids. The two groups of monkeys (Old World and New World) separated from the anthropoid branch fairly early (more than 20 million years ago), leaving the clade called *hominoids*. (Notice the second "o" in that word—it's going to disappear in the next paragraph, and the new word will mean something else.) The hominoids are known as the apes, that is, the gibbons, orangutans, gorillas, chimpanzees, and humans; they are characterized by a relatively large brain and no tail. The large brain may be why they are so behaviorally diverse.

Scientists have found fossils of approximately 20 species that are more similar to humans than to the rest of the hominoids. Along with humans, these species (and probably some we have yet to discover) are called *hominids*, a subset of the hominoids, with a name that lacks the second "o." The earliest hominid lived more than 4 million years ago, perhaps much earlier. Humans are the only living representatives of the hominid branch, but the fossil hominids tell our story.

What traits do the hominids share? Some of the earliest hominids—fossils more than 4 million years old—already show evidence of bipedalism, the tendency to walk upright, something that the apes are not very good at. You may wonder how we can discover behavior from a fossil. In the case of bipedal locomotion, a upright stance is associated with changes in the location of the spinal cord relative to the skull. The spinal cord enters the skull through a hole called the *foramen magnum*. If that hole is directly under the skull, as it is in hominids, rather than to the rear of the skull, as it is in chimpanzees, it means that the hole-under-the-skull animal could hold its head directly over its body. That saves lots of energy (a head is remarkably heavy) and is a hallmark of efficient bipedal locomotion. Other traits that we like to think about, such as our large brain, came along later in hominid evolution. We share that with some, but not all hominids. The brains of early hominids were 400–450 cm^3, one-third the size of ours.

This tells us that bipedalism—that is, walking on the ground, and spending decreasing amounts of time in the trees, with all the accompanying ecological changes—happened early in human evolution and did not require a big brain. What caused our ancestors to leave the trees? We don't know for sure, but some scientists speculate that the climate was changing, and forests were diminishing in size, leaving the grasslands called savannahs. The more hominid fossils we discover, the more we realize that the evolution of bipedalism was not a smoothly linear process. Among other things, it is clear that early hominids lived in environments that were mixtures of forests and savannahs. Full adaptation to bipedalism includes not only the shift in the foramen magnum discussed above, but also a shift in pelvic orientation, a downward shift in center of gravity, straighter legs, and a foot that is specialized for locomotion. Although the apes can move bipedally, their legs are more flexed, they have a higher center of gravity, and their feet are a bit more general purpose. Apes can use their legs for grasping more than we can, but they are not as good at transferring force from the leg muscle to the ground, that is, locomotion. On the whole, apes can move bipedally, but they are less well adapted to that form of locomotion; bipedalism costs them more energy than it does us.

Between 2 and 4 million years ago, hominid diversity increased (Figure 5.1). Many hominids of that time belong to a genus named *Australopithecus* ("southern ape"). This genus includes the fossil that its discoverers (Donald Johanson and his colleagues) nicknamed "Lucy" (after they celebrated their discovery by listening to the Beatles' song *Lucy in the Sky with Diamonds*). Lucy is a remarkable fossil because it is 40 percent complete. "Lucy" was found in Ethiopia; she lived over 3 million years ago and was bipedal. She also had an ape-sized brain, not a human-sized one.

Beginning 2.4 million years ago, some new traits appeared among hominid fossils. The primary difference between these hominids and the australopithicines (i.e., *Australopithecus*) is the substantially larger brain. Anthropologists place these large-brained hominids in the genus *Homo* ("man"). Fossils of *H. habilis* ("handy man") are the oldest in the genus and have been found with stone tools, but at this point, their tools may be more abundant, or at least easier to find, than their fossils. At 900 cm^3, *H. ergaster*, sometimes included in *H. erectus,* had an even larger brain than *H. habilis*. *H. ergaster* lived 1.5–1.9 million years ago.

The last of our fellow hominids to become extinct (30,000 years ago) was perhaps the poster child of human evolution—certainly the most maligned of hominids—*Homo neanderthalensis* (Figure 5.2). The Neanderthal man, whose fossils were first found in the Neander Valley of Germany, was heavier than *H. sapiens,* with thick bones, a prominent brow, and a brain every bit as large as ours. We find this relative of ours intriguing. Neanderthals were not our immediate ancestors; they

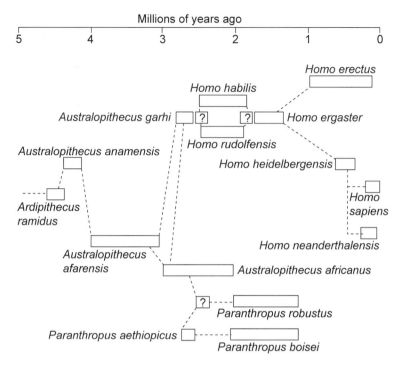

Figure 5.1 Hominid timeline. Hominid lineages that would eventually go extinct coexisted with ancestors of humans. (*Jeff Dixon*)

were in a different lineage than we are. They fashioned tools from wood and stone. What would the world be like if we shared it with *H. neanderthalensis?* What happened to them 30,000 years ago that caused their disappearance? We don't know for sure, but there is evidence that *H. sapiens,* while having a similar-sized brain (actually, a bit smaller), might have used that brain in a more sophisticated and complex fashion. *Homo sapiens* of that time (36,000 years ago) was probably capable of symbolic thought. We know they left a record of lovely cave paintings; they may have even used language. With language and the ability to plan, trading networks can be established. In short, *H. sapiens* might have been a tough competitor.

Our large brains do indeed allow us to do complex things. Why would natural selection favor, say, the ability to do calculus? It probably didn't. But as hominid social groups formed—all the better to find food and detect predators—it became increasingly advantageous for group members to remember one another and to devise ways to cooperate and

Figure 5.2 Michael Anderson's reconstructions of the Neanderthal show first the skull form, then facial and neck musculature, and finally a lifelike form with skin, hair, and eyes. (*Photo credit: Melanie Brigockas*)

communicate, not to mention, deceive. Indeed, most social mammals have larger brains than related solitary species. Language itself could favor larger and larger brains, as it became an increasingly important survival tool. Perhaps language and intelligence were appealing traits in a mate.

Scientists are working to understand more about human evolution. Molecular genetics may eventually reveal the relationship patterns among groups of humans and may explain how the world was colonized by those early Africans.

Meanwhile, before you feel too pleased about your complex brain, your skill at language, and your general good fortune to be in the *Homo sapiens* lineage, remember that there are over 350,000 species of beetles alone. Right now, some of them are eating your lunch—or what might have been your lunch. Remember, too, that there are animals that communicate volumes using smells, sounds, the smallest curl of a lip—signals that we do not—and cannot—perceive.

Evolution does not tell us that we are at the top of some ladder, looking down on the various vermin and weeds below. Instead, evolution tells us that we are honored to be part of the colorful, ever-changing fabric of all life. The same forces that produced bees, trees, bears, fish, and other forms of life produced us. But we are also different, for no other animal takes time to think about and help the sick and dying to the extent we do, or tries to save those hurt in floods, fires, and tsunamis. And if we are, indeed, the only species that can contemplate that fact, then how much more should we care for our Earth and fellow organisms.

We have named ourselves *Homo sapiens*—"wise man." We are wise enough to contemplate our own origin, our relationship to other living things. How we care for that living web will put all our resources of heart and mind to the test. It will tell us if we are indeed wise.

SUMMARY

The biggest breakthrough in the history of biology was Charles Darwin's theory of evolution by natural selection. Although the impact of other intellectual giants of the 19th century—Karl Marx's (1818–1883) materialistic theory of the history of society, and Sigmund Freud's (1856–1939) attribution of human behaviors to influences over which we have little control—have faded, Darwin's idea continues to tell us not only about the history of living things, but also about contemporary issues. Darwinian concepts, when misunderstood and misapplied, have given rise to social problems such as Social Darwinism and eugenics. When used rightly, however, those concepts offer an understanding of nature and solutions to such major challenges as antibiotic resistance, pesticide resistance, and disease. At some level, every discovery in biology and medicine rests on Darwin's idea.

APPENDIX 1

THE GEOLOGICAL TIMESCALE

Age (millions of years ago)*	Era	Period	Events
0.1	Cenozoic		End of last Ice Age
			First human civilizations appear with cities, agriculture, and the domestication of animals
			Human activities reduce biological diversity to the lowest levels since the Mesozoic era (65 million years ago)
1.8		Quaternary	Many Ice Ages
			Extinction of some plants and large mammals
			Humans appear
5			Climate cools
			North and South America join at Isthmus of Panama
			Forests decline as grasslands and deserts expand
			Diversification of grazing and carnivorous mammals
			Large mammals roam the earth
			Apelike ancestors of humans appear in Africa
			Flowering plants dominate land

(continued)

(*continued*)

Age (millions of years ago)*	Era	Period	Events
24			Many mountains form
			Primates diversify
			Great diversification of grazing mammals and songbirds
33			Rise of Alps and Himalayas
			Volcanic activity in Rockies
			Forests spread as flowering plants diversify
			Apes appear
55			Climate warms
			Flowering plants dominate land
			First primates appear
			Modern birds and mammals continue to diversify
65		Tertiary	Climate mild
			Diversification of mammals, birds, and pollinating insects
			Flowering plants and primitive mammals diversify
			Gymnosperms decline

The Cretaceous-Tertiary (K-T) mass extinction eliminates more than 60 percent of all species. This is the most famous mass extinction, not because of its magnitude (the Permian mass extinction wiped out many more species), but because of its most famous victims: dinosaurs. Crocodiles, turtles, birds, lizards, and mammals are relatively unaffected.

Age (millions of years ago)*	Era	Period	Events
144	Mesozoic	Cretaceous	Continents separate
			Formation of large inland seas and swamps
			Flowering plants, social insects, and primitive mammals appear
			Many groups of organisms, including toothed birds and most dinosaur lineages, become extinct (Cambrian extinctions)
206		Jurassic	Mountains and inland seas continue to form
			Gymnosperms dominate land

Age (millions of years ago)*	Era	Period	Events
			Dinosaurs are abundant
			Toothed birds and large, specialized dinosaurs appear

The Triassic mass extinction eradicates about 20 percent of all species, thereby opening niches that allow the diversification of dinosaurs.

Age (millions of years ago)*	Era	Period	Events
251		Triassic	Many mountains form; deserts spread
			Gymnosperms dominate land
			Frogs and turtles evolve
			Radiation of dinosaurs, birds, and mammals
290	Paleozoic	Permian	Continents rise and merge as Pangaea
			Conifers diversify as cycads appear
			Modern insects appear
			Radiation of reptiles
			Many invertebrates and vertebrates become extinct (Permian extinctions)

The Permian mass extinction kills more than 90 percent of all species. The Permian mass extinction is the most devastating mass extinction.

Age (millions of years ago)*	Era	Period	Events
354		Carboniferous	Land is swampy; climate is mild
			Amphibians dominate
			Extensive forests of vascular plants dominate land
			Reptiles appear and become the first vertebrates to have a life cycle fully independent of water
			First seed-plants appear
408		Devonian	Diversification of bony fishes
			Amphibians and insects appear

The Devonian mass extinction eradicates as many as two-thirds of all species. Most of these extinctions are of marine species; there are far fewer losses on land.

(*continued*)

(*continued*)

Age (millions of years ago)*	Era	Period	Events
439		Silurian	Most continents remain covered by seas
			Land colonized by vascular plants and arthropods
			First jawed fishes appear
495		Ordovician	Most continents are covered by seas
			Marine algae abundant
			Origin of plants

The Ordovician-Cambrian mass extinction eliminates most marine species; many groups lose more than half of their species. We know relatively little about the cause and impact of this mass extinction because most animals were soft-bodied, and therefore unlikely to have become fossils.

Age	Era	Period	Events
543		Cambrian	Origin of most animal phyla (Cambrian explosion)
610	Precambrian		Algae and soft-bodied invertebrates diversify
700			Oldest known animal fossils
1700			Oldest known eukaryotic fossils
2500			Oxygen begins accumulating in the atmosphere
3500			Oldest known fossils (prokaryotes)
4600			Origin of Earth

* Although these dates have an accuracy of approximately ±1%, they continue to change as geologists examine more rocks and refine dating methods.

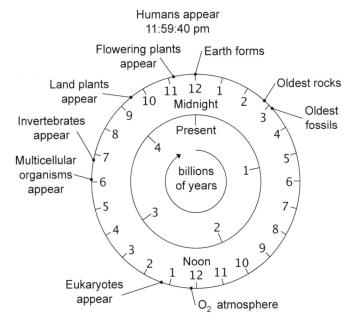

Figure A1.1 Timescale for the evolution of life on Earth. If we think of the history of the Earth as a 24-hour day, starting at midnight, then eukaryotes do not appear until early afternoon. Most of the diversity of life appears during the evening and night; humans appear at the end of the day—just a few seconds before midnight. (*Jeff Dixon*)

APPENDIX 2

LEGAL DECISIONS INVOLVING THE TEACHING OF EVOLUTION AND CREATIONISM IN PUBLIC SCHOOLS

1925	*State of Tennessee v. John Thomas Scopes*	In the original "Trial of the Century" (see "Evolution in the Courtroom," Chapter 2), coach and substitute science-teacher John Scopes was convicted of the misdemeanor of teaching human evolution in a public school in Tennessee. Scopes' trial, which William Jennings Bryan described as "a duel to the death" between evolution and Christianity, remains the most famous event in the history of the evolution–creationism controversy (see Chapter 2, Evolution in the Courtroom). In 1960, the "Scopes Monkey Trial" also provided a framework for the Academy Award winning movie, *Inherit the Wind.*
1927	*John Thomas Scopes v. State of Tennessee*	The Tennessee Supreme Court upheld the constitutionality of a Tennessee law forbidding the teaching of human evolution, but urged that Scopes' conviction be set aside. This decision ended the legal issues associated with the Scopes Trial, and the ban on teaching human evolution in Tennessee, Mississippi, and Arkansas remained unchallenged for more than 40 years.
1968	*Epperson v. Arkansas*	The U.S. Supreme Court struck down an Arkansas law making it illegal to teach human evolution. As a result of this decision, all laws banning the teaching of human evolution in public schools were overturned by 1970.
1972	*Willoughby v. Stever*	The D.C. Circuit Court of Appeals ruled that government agencies such as the National Science

(*continued*)

(*continued*)

		Foundation can use tax money to disseminate scientific findings, including evolution.
1973	*Wright v. Houston Independent School District*	The 5th Circuit Court of Appeals ruled that (1) the teaching of evolution does not establish religion, (2) there is no legitimate state interest in protecting particular religions from scientific information "distasteful to them," and (3) the free exercise of religion is not accompanied by a right to be shielded from scientific findings incompatible with one's beliefs.
1975	*Daniel v. Waters*	The 6th Circuit Court of Appeals overturned the Tennessee law requiring equal emphasis on evolution and the Genesis version of creation.
1977	*Hendren v. Campbell*	The County Court in Marion, Indiana, ruled that it is unconstitutional for a public school to adopt creationism-based biology books because these books advance a specific religious point of view.
1980	*Crowley v. Smithsonian Institution*	The D.C. Circuit Court of Appeals ruled that the federal government can fund public exhibits that promote evolution. The government is not required to provide money to promote creationism.
1982	*McLean v. Arkansas Board of Education*	An Arkansas federal district court ruled that creation science has no scientific merit or educational value as science. Laws requiring equal time for "creation science" are unconstitutional.
1987	*Edwards v. Aguillard*	The U.S. Supreme Court overturned the Louisiana law requiring public schools that teach evolution to also teach "creation science," noting that such a law advances religious doctrine and therefore violates the First Amendment's establishment of religion clause.
1990	*Webster v. New Lenox School District #122*	The 7th Circuit Court of Appeals ruled that a teacher does not have a First Amendment right to teach creationism in a public school. A school district can ban a teacher from teaching creationism.
1994	*Peloza v. Capistrano Unified School District*	The 9th Circuit Court of Appeals ruled that evolution is not a religion and that a school can require a biology teacher to teach evolution.
1996	*Hellend v. South Bend Community School Corporation*	The 7th Circuit Court of Appeals ruled that a school must direct a teacher to refrain from expressions of religious viewpoints (including creationism) in the classroom.

1999	*Freiler v. Tangipahoa Parish Board of Education*	The 5th Circuit Court of Appeals ruled that it is unlawful to require teachers to read aloud a disclaimer stating that the biblical view of creationism is the only concept from which students are not to be dissuaded. Such disclaimers are "intended to protect and maintain a particular religious viewpoint."
2000	*LeVake v. Independent School District #656*	A Minnesota state court ruled that a public school teacher's right to free speech as a citizen does not permit the teacher to teach a class in a manner that circumvents the prescribed course curriculum established by the school board. Refusing to allow a teacher to teach the alleged "evidence against evolution" does not violate the free-speech rights of a teacher.
2001	*Moeller v. Schrenko*	The Georgia Court of Appeals ruled that using a biology textbook that states creationism is not science does not violate the Establishment or the Free Exercise Clauses of the Constitution.
2005	*Selman et al. v. Cobb County School District*	The U.S. District Court for the Northern District of Georgia ruled that it is unconstitutional to paste stickers claiming that, among other things, "evolution is a theory, not a fact," into science textbooks. Such stickers convey "a message of endorsement of religion" and "aid the belief of Christian fundamentalists and creationists." However, in 2006, this decision was vacated on appeal by the 11th Circuit Court of Appeals, which remanded the case for further evidential proceedings.
2005	*Kitzmiller et al. v. Dover Area School District*	The U.S. District Court for the Middle District of Pennsylvania ruled that (1) "the overwhelming evidence . . . established that intelligent design (ID) is a religious view, a mere re-labeling of creationism, and not a scientific theory," and, instead, is nothing more than creationism in disguise, (2) the advocates of ID wanted to "change the ground rules of science to make room for religion, and (3) "ID is not supported by any peer-reviewed research, data, or publications." The judge also noted the "breathtaking inanity" of the school board's policy and the board's "striking ignorance" of ID and made the following point: "It is ironic that several of [the members of the School Board], who so staunchly and proudly touted their religious convictions in public, would time and again lie to cover their tracks and disguise the real purpose behind the ID Policy."

Figure A 2.1 In 1960, *Inherit the Wind* used the Scopes Trial to examine intolerance and zealotry. Although the movie was not a historically accurate portrayal of the trial, many people mistook it as a documentary. (*Randy Moore*)

FUTURE PROSPECTS

Although U.S. courts have struck down all attempts to introduce creationism into science classes of public schools, most politicians continue to endorse creationism. Some political leasers have linked the teaching of evolution with school violence, and a state legislator in Louisiana introduced a bill blaming evolution for racism (in fact, both evolution and creationism have often been used to justify racism). Do not expect the evolution–creationism controversy to end.

REFERENCES

Crowley v. Smithsonian Institution, 636 F.2d 738 (D.C. Cir. 1980).

Daniel v. Waters, 515 F. 2d 485 (6th Cir. 1975).

Edwards v. Aguillard, 482 U.S. 578 (1987).

Epperson v. Arkansas, 393 U.S. 97 (1968).

Freiler v. Tangipahoa Parish Board of Education, 185 F.3d 337 (5th Cir. 1999), *cert. denied*, 530 U.S. 1251 (2000).

Hellend v. South Bend Community School Corporation, 93 F.3d 327 (7th Cir. 1996), *cert. denied*, 519 U.S. 1092 (1997).

Hendren v. Campbell, Superior Court No. 5, Marion County, Indiana, April 14, 1977.

Kitzmiller et al. v. Dover Area School District, No. 4:04CV02688 (M. D. Pa.), December 20, 2005.

LeVake v. Independent School District, #656, 625 N.W.2d 502 (MN Ct of Appeal 2000), *cert. denied*, 534 U.S. 1081 (2002).

McLean v. Arkansas Board of Education, 529 F. Supp. 1255, (E.D. Ark. 1982).

Moeller v. Schrenko, 554 S.E.2d 198 (GA Ct. of Appeal 2001).

Peloza v. Capistrano Unified School District, 37 F.3d 517 (9th Cir. 1994).

Scopes v. State of Tennessee, 289 S.W. 363 (Tenn. 1927).

Selman et al. v. Cobb County School District, No. 1:02CV2325 (N. D. Ga. filed August 21, 2002, decided January 13, 2005).

State of Tennessee v. John Thomas Scopes (1925), reprinted in *The World's Most Famous Court Trial, State of Tennessee v. John Thomas Scopes*. New York: Da Capo Press (1971).

Webster v. New Lenox School District, #122, 917 F. 2d 1004 (7th Cir. 1990).

Willoughby v. Stever, Civil Action No. 1574–72 (D.D.C. August 25, 1972), *aff'd mem.*, 504 F.2d 271 (D.C.Cir. 1974), *cert. denied*, 420 U.S. 927 (1975).

Wright v. Houston Independent School District, 366 F. Supp. 1208 (S.D.Tex. 1972), *aff'd*, 486 F.2d 137 (5th Cir. 1973), *cert. denied sub. nom.* Brown v. Houston Independent School District, 417 U.S. 969 (1974).

APPENDIX 3

A TIMELINE FOR EVOLUTIONARY THOUGHT

~440 B.C.   Greek philosopher Empedocles claims that the universe and everything in it is gradually changing (e.g., "Many races of living creatures must have been unable to continue their breed . . .").

~360 B.C.   Socrates' student Plato proposes that nature consists of transcendent, ideal forms, and that variations from these forms are illusions.

~310 B.C.   Plato's student Aristotle proposes that life consists not just of a listing of ideal forms, but that each form of life has its own static position which reflects its degree of "perfection" in a ladder of nature. The so-called "Great Chain of Being" (Figure 4.5) originally extended from worms (which occupied the lowest link) to humans (which occupied the highest link). Christians later extended the chain to angels, archangels, and God.

392   The Christianization of the Roman Empire is nearly complete when Emperor Theodosius makes Christianity the only approved religion. As a result, several western ideas crucial to what would become biology can be traced to Genesis, including the idea that all life forms are the direct, intentional, and sudden creation of God, and that life has not changed much since the Creation.

1541   Martin Luther (1483–1546) claims that creation occurred in 3961 B.C. Virtually all subsequent biblical chronologies make similar claims.

1644   John Lightfoot, Vice Chancellor of the University of Cambridge, claims that creation occurred at 9 a.m. on Sunday, October 23, 4004 B.C.

1650–54   Irish archbishop and scholar James Ussher uses the Bible and other ancient texts to claim that creation occurred in 4004 B.C. Ussher started with the known dates of the reign of

Nebuchadnezzar II and subtracted the life spans of the Bible's patriarchs. These calculations, combined with several assumptions (e.g., that creation occurred on an equinox or solstice, that the presence of a ripe fruit meant that creation occurred in the autumn), produced Ussher's claim. Ussher's *Annals of the World* described the adventures of Abraham, Julius Caesar, Alexander the Great, and other historical figures.

1660   Italian physician Francesco Redi strikes a blow against spontaneous generation by showing that maggots appear when meat is exposed to open air, but do not occur in meat that is isolated from flies.

1669   Nicolaus Steno's 78-page *De Solido* opens geology to the dimensions of time by showing that Earth's history is written in its rocks.

1701   John Ray writes that organisms have always been the same, lived in the same places, and done the same things as when they were first created.

1735   Linnaeus begins publishing *Systema Naturae*, which establishes the foundation of systematic biology. Linnaeus places every plant and animal into its own unique and static place in the divinely created order of nature. His system is later adapted to evolutionary interpretations.

1739   The first fossils found in the United States are gathered by Charles de Longueuil from a marsh at what is now Big Bone Lick State Park in Kentucky.

1744   Jean-Baptist Lamarck is born in France. In 1809, Lamarck will propose the first testable hypothesis for evolution.

1749   Buffon (Georges-Louis Leclerc) suggests that Earth might be more than 70,000 years old and rejects claims that life reflects a divine plan of creation. Buffon begins publishing *Histoire Naturelle*, which eventually appears in 44 volumes (the last eight were published, using Buffon's notes, after his death in 1788). *Histoire Naturelle* is one of the most influential works in the history of biology.

1753   Linnaeus introduces the binomial system of nomenclature in *Species Plantarum.*

1758   Linnaeus uses Volume 1 of the 10th edition of *Systema Naturae* to spell out his binomial system of classification. In this book, Linnaeus introduces to biology, and defines, the terms *Mammalia, Primates,* and *Homo sapiens,* among others. In his publications, Linnaeus describes some 7,700 species of plants and 440 species of animals, which comprise most of the known species in Europe at the time.

1766   In the 12th edition of *Systema Naturae*, Linnaeus abandons his earlier claim that new species do not form.

              Buffon argues that similar species share a common ancestor and that their geographic distribution provides clues about the species' history.

1776   Adam Smith's *Wealth of Nations* provides a blueprint for how free enterprise can produce economic growth. Smith's *laissez faire* depicts an "invisible hand" that guides individuals to act for personal gain that produces economic growth. Just as Smith tried to understand "the nature and causes of wealth," Thomas Malthus would later try to determine "the nature and causes of poverty." Charles Darwin will apply these principles to nature itself.

1785   Scottish farmer and geologist James Hutton presents uniformitarianism in two papers read to the Royal Society of Edinburgh. Criticisms prompt Hutton to expand his ideas into *Theory of the Earth, with Proofs and Illustrations*. Hutton claims that all creatures except humans are ancient, and that all of Earth's changes have been purposeful (i.e., the world has been constantly replenished for human's use).

1794–96   Erasmus Darwin, the grandfather of Charles Darwin, publishes the 200,000-word *Zoonomia*, in which he suggests evolution has occurred, but offers no mechanism.

1795   James Hutton publishes his 2-volume book, *Theory of the Earth*, in which he notes change, not stasis, explains geology and that "in examining things present we have data from which to reason with regard to what has been." Hutton had planned a third volume, but he never completed it. Hutton regarded the age of the Earth as beyond comprehension, as noted in his most famous line, "we find no vestige of a beginning—no prospect of an end." Hutton's ideas are later developed by Charles Lyell and become the basis of *uniformitarianism.*

1796   Georges Cuvier confirms extinctions and claims that they have been caused by God-directed catastrophes. Cuvier, who established the discipline of paleontology, vehemently rejects evolution all his life.

              Thomas Jefferson hires William Clark (later of Lewis and Clark fame) to collect the remains of extinct mastodons in Kentucky. Clark would later search for relict populations of these mastodons in the American West, but would not find them. The idea that extinct creatures are alive elsewhere continues to resurface for centuries (e.g., the Loch Ness plesiosaur, Conan Doyle's classic *Lost World*).

JAMES HUTTON
M.D., F.R.S.E.,
1726 – 1797.

THE FOUNDER OF MODERN
GEOLOGY

Figure A 3.1 James Hutton was intrigued by the 73-mile-long Hadrian's Wall (upper photo), which was built by Romans in Britain around 122 A.D. "to separate the Romans from the Barbarians." The wall, named after Roman Emperor Hadrian (who ordered

1797   Charles Lyell is born in the same year that Hutton dies at age 71. Lyell's *Principles of Geology* will later have a tremendous influence on Charles Darwin's views of nature.

1798   In *Essay on a Principle of Population* (published anonymously), philosopher and economist Thomas Malthus refutes the claim that social reform can bring happiness to all, instead proposing that the sizes of populations increase geometrically and therefore outstrip the availability of food. Malthus' reasoning that suffering is a basic part of nature undermines natural theology's attempts to promote a benevolent deity. Malthus' starkly simple argument that the competition for limited resources must produce a "struggle for existence" is critically important for Darwin's idea that evolution occurs by natural selection. Malthus later becomes a leading economist and is elected a Fellow of the Royal Society in 1819.

1802   William Paley publishes his bestseller *Natural Theology, Or, Evidence of the Existence and Attributes of the Deity Collected from the Appearances of Nature,* which establishes his argument for design. Paley claims that God created organisms perfectly adapted to their environment. Paley's book, which mentions neither miracles nor the Bible, remains the greatest exposition of natural theology. Ironically, natural theology supplants some earlier theological ideas and sets the stage for the creationism debate.

Hutton's friend John Playfair (1748–1819) publishes *Illustrations of the Huttonian Theory of the Earth,* a book in which Hutton's principle of uniformitarianism first reaches a wide audience. In 1831, Charles Darwin will take *Illustrations* aboard the *Beagle.*

Erasmus Darwin dies at age 70.

1807   The Geological Society of London is founded. In 1859, it will award Charles Darwin its highest honor.

1809   Jean Baptiste Lamarck argues for the inheritance of acquired traits and that evolution is driven by organisms' needs as they strive to fulfill their way of life. Lamarck's ideas form the first testable theory—albeit inaccurate—to explain how organisms evolve. Lamarck's ideas were rejected by other scientists.

Figure A 3.1 *(continued)* its construction), marked the northern limits of the Roman Empire. Rocks in the wall were almost 1,700 years old, but had not weathered at all. However, nearby mountains had eroded a great deal. Hutton concluded from this and many other observations that Earth must be very old. A plaque at Hutton's grave (lower photo) notes that he was "The Founder of Modern Geology." (*Janice Moore*)

Figure A 3.2 In 2000, Charles Darwin replaced fellow Victorian Charles Dickens on the British 10-pound note. (*Randy Moore*)

Charles Robert Darwin is born into affluence in Shrewsbury, Shropshire, England. Darwin shares his birthdate with Abraham Lincoln, both of whom will later appear on the currency of their country (Figure A3.2).

Cuvier, who resolutely treats Genesis as historical, asks, "would there not also be some glory for man to know how to burst the limits of time, and . . . to recover the history of the world, and the succession of events that preceded the birth of the human species?"

1822   Mary Ann Mantell and her husband Gideon discover the first dinosaur bones in England's Tilgate Forest. The dinosaur is later named *Iguanodon.*

1823   Alfred Russel Wallace is born in Usk, Wales. He will later be the co-discoverer (with Charles Darwin) of the theory of evolution by natural selection. In 2006, Wallace's admirers will erect a monument to Wallace outside the church where Wallace was baptized and close to the cottage in which he was born.

1827   Charles Darwin's first scientific paper, a study of fertilization in the seaweed-like bryozoan *Flustra,* is read to the Plinian Society.

1828   Charles Darwin begins 3 years of study at Cambridge, staying in the same rooms formerly occupied by William Paley. While there, he has conventional beliefs, but a notable lack of religious zeal. Darwin attends botany lectures and goes on plant-collecting expeditions led by John Henslow, who later helps Darwin secure a position on the HMS *Beagle.* Darwin reads John Herschel's *Introduction to the Study of Natural Philosophy,* which "stirred up in [Darwin] a burning zeal to add even the most humble contribution to the noble structure of Natural Science." In the introduction of *On the Origin of Species*, Darwin refers to Herschel as the great philosopher who coined the phrase "mystery of mysteries" to describe the change of Earth's species over time. Darwin would later be buried beside Herschel in Westminster Abbey.

Charles Lyell, working in Sicily, finds raised seabeds "700 feet and more" above sea level. During the same trip, Lyell visits the ruins of the Temple of Serapis, whose columns are marked by bands of rock-boring mollusks (Figure 1.5). When Lyell later places a sketch of the columns in the frontispiece of Volume 1 of his *Principles*, they become an icon for uniformitarianism.

1829   Lamarck dies and is buried in a rented grave. Although Cuvier's eulogy ridicules Lamarck's "fanciful conceptions" and indulgent "imagination," Lamarck's daughter provides a poignant and prophetic epitaph, "Posterity will remember you."

The Linnean Society is formed to preserve the collections and books of Linnaeus. In 1858, the Linnean Society will host the first public announcement of Darwin's and Wallace's theory of evolution by natural selection.

1830–33   Charles Lyell's 3-volume *Principles of Geology, being an Attempt to Explain the Earth's Surface by Reference to Causes Now in Operation* outlines the story of an old Earth in which species are constantly becoming extinct as others emerge. Lyell argues that Earth's features have been formed by ordinary forces still in operation. *Principles*, a name chosen to deliberately echo Newton's *Principia*, is an immediate success. Lyell's book will later guide Darwin's thinking about nature; as Darwin noted, "I always feel as if my books came half out of Lyell's brain ... I have always thought that the great merit of the Principles was that it altered the whole tone of one's mind."

Charles Darwin graduates from Christ's College, Cambridge.

1831   Thanks to a recommendation from John Henslow, Charles Darwin leaves on a 5-year cruise aboard the HMS *Beagle*. Before the voyage, Captain FitzRoy gives Darwin a copy of the first volume of Lyell's *Principles of Geology*, but Darwin's favorite book on the *Beagle* is Milton's *Paradise Lost*. Darwin later describes his voyage aboard the *Beagle* as "the most important event in my life." He would not publish *On the Origin of Species* until 28 years later.

1832   During the *Beagle*'s stop at Montevideo, Darwin receives the second volume of Lyell's *Principles of Geology*. This volume, which is best known for Lyell's attack on Lamarck's ideas about evolution, reflects Lyell's interest in "changes in the organic world now in progress."

Philosopher and scientist William Whewell introduces geologists to the terms *uniformitarianism* and *catastrophism*.

1833   The third volume of Lyell's *Principles of Geology* is published. Lyell will spend most of the rest of his life revising *Principles*, which provides most of his income.

1834   Political economist Thomas Malthus, whose *Essay on Population* would have a major impact on Darwin, Wallace, and others, dies while Darwin is at sea aboard the *Beagle*.

1835   Darwin spends a month at the Galápagos Islands.

Darwin experiences Hutton's and Lyell's ideas in action when he watches a volcano erupt in Chile and, a month later, feels an earthquake in Valdivia.

1836   Charles Darwin returns to England. He never again leaves England.

Charles Darwin meets Charles Lyell at a dinner at Lyell's home.

1837   Darwin begins writing the first of several notebooks about the "species question." Darwin knows that species change, but can't explain how. Darwin reads a paper to the Geological Society of London about the coastal uplift in Chile, one of the biggest discoveries during his time aboard the *Beagle*. Soon thereafter, Darwin is elected a Fellow of the Geological Society of London.

1838   Darwin reads Malthus' *Essay on Population* and discovers his theory of natural selection. To Darwin, natural selection is the force that constantly adjusts the traits of future generations.

Lyell publishes *Elements of Geology*, the first modern textbook of geology.

1839   Darwin marries his first cousin Emma Wedgewood, a wealthy woman interested in the arts and travel; she had even seen the Temple of Serapis, which Lyell had included in his *Principles*.

Darwin publishes *Journal of Researches*, which attacks slavery and describes Darwin's joy in observing Earth's geological features. Publication was delayed while FitzRoy completed his part of the three-book volume. To FitzRoy's chagrin, Darwin's book is the more popular, and it is quickly republished on its own as the *Voyage of the Beagle*. Like Lyell, Darwin receives acclaim as a scientist and writer.

Darwin is elected a Fellow of the Zoological Society and the Royal Society.

1842   Darwin completes a 35-page handwritten summary of natural selection.

Darwin publishes *The Structure and Distribution of Coral Reefs*, the first volume of his geological trilogy (other volumes will include *Volcanic Islands* in 1844, and *Geological Observations on South American* in 1846). Through 1846, Darwin publishes 19 papers (or notices) about geological topics, after which his day-to-day interests shift to biology.

1844   Darwin expands his earlier summary of natural selection into a 231-page essay that rejects the fixity of species.

Robert Chambers anonymously publishes his best-selling *Vestiges of the Natural History of Creation*, in which he claims that Earth was not specifically created by God, but formed by laws that expressed the Creator's will.

Darwin publishes *Geological Observations on Volcanic Islands*.

1845   John Murray publishes the second edition of Charles Darwin's *Journal of Researches*. By this time, Darwin had become "an unbeliever in every thing beyond [my] own reason." Fourteen years later, Murray will publish Darwin's *On the Origin of Species*. (Today, John Murray Publishers Ltd. continues to be run by the Murray family.)

1848   Lyell, the leading geologist of his time, is knighted.

1850   Alfred, Lord Tennyson, one of the most famed poets of the Victorian age, becomes poet laureate and publishes *In Memoriam*, which he has worked on for 17 years. In the 131-stanza poem, Tennyson describes a growing sense that nature is characterized by suffering and death, and that it gives no comfort to people

> Who trusted God was live indeed
> And love creation's final law-
> Tho' Nature, red in tooth and claw
> With ravine shriek'd against his creed.

1851   The death of 10-year-old Annie Darwin, Charles' "little angel" and favorite child, removes the last vestiges of religious faith from Darwin. Neither Charles nor Emma attend Annie's funeral.

1854   Alfred Russel Wallace departs for the Malay Archipelago. From there, in 1858, Wallace will send to Charles Darwin a letter and manuscript that prompts Darwin to write and publish *On the Origin of Species*.

1856   Darwin begins work on his "big book," which will become the monumental *On the Origin of Species by Means of Natural Selection*.

Charles Lyell visits Down House and urges Darwin to publish his theory. Darwin responds that "I rather hate the idea of

Figure A 3.3 The death of Annie Darwin in 1851 changed her famous father's views of Christianity. (*Reprinted with permission of HarperCollins Publishers*)

writing for priority, yet I certainly should be vexed if anyone were to publish my doctrine before me."

Neanderthal man (*Homo neanderthalensis*) is discovered in the Neander Valley in Germany. This is the first recognized fossil human form.

1858   Charles Darwin receives a letter from 35-year-old Alfred Russel Wallace containing a manuscript entitled "On the Tendency of Varieties to Depart Indefinitely from the Original Type" that describes Wallace's ideas about natural selection. Wallace's ideas are virtually identical to those of Darwin. As per Wallace's request, Darwin sends Wallace's manuscript to Charles Lyell. Wallace later rejects Darwin's claim that natural selection also operates on humans.

Wallace's paper, a letter from Charles Darwin to Asa Gray, and an extract from Darwin's essay written in 1844 are presented to the Linnean Society by Joseph Hooker and Charles Lyell. Darwin does not attend the presentation because his son Charles Waring Darwin had died 2 days earlier of scarlet fever. The presentation attracts little interest and produces no controversy; at year's end, the Linnean Society president reports that "The year . . . has not, indeed, been marked by any of those striking discoveries which at once revolutionize, so to speak, the department of science on which they bear."

Rudolf Virchow shows that every living cell comes from a preexisting cell.

1859 Charles Lyell suggests to publisher John Murray that Murray might want to publish Charles Darwin's "important new work." Murray agrees.

Charles Darwin finishes the last chapter of *On the Origin of Species*, and later in the year, corrects proofs.

The Geological Society of London awards Darwin with the Wollaston Medal, its highest honor.

Darwin tells Lyell that Lamarck's views about evolution are "wretched . . . and from which I gained nothing."

All 1,250 copies of Darwin's 502-page *On the Origin of Species by Means of Natural Selection, or the Preservation of Favoured Races in the Struggle for Life* sell on the first day after publication. Darwin models *Origin* after Lyell's *Principles of Geology*. Darwin takes a vacation and waits for the upcoming controversy. *Origin* will eventually go through six editions, the last of which will appear in 1872. During Darwin's lifetime, British printings alone sell more than 27,000 copies. A few days after the publication of *On the Origin of Species*, Darwin tells Lyell that he believes that his theory will explain how man evolved as a thinking being.

1860 The second edition of *On the Origin of Species* is published.

Harvard botanist Asa Gray's review of Darwin's *On the Origin of Species* notes that Darwin's theory is not necessarily atheistic. Darwin describes the review as "by far the best which I have read."

The American Academy of Arts and Sciences convenes a special meeting to discuss Darwin's ideas.

Harvard biologist Louis Agassiz denounces Darwin's theory as a collection of "mere guesses" that is "a scientific mistake, untrue in its facts, unscientific in its methods, and mischievous in its tendency."

Thomas Huxley, Darwin's fiercest defender, has a famous debate with Samuel Wilberforce, the Bishop of Oxford. The debate is the first exhibit of open resistance to the church's authority regarding human origins. The debate is attended by Robert FitzRoy, the man who provided Darwin with the vehicle for his conclusions. FitzRoy, dressed in a rear admiral's uniform, stands and waves a Bible over his head as he tells listeners that he regrets the publication of Darwin's book. Few pay attention to FitzRoy's comments.

Asa Gray's widely read article in *Atlantic Monthly* claims that Darwin's ideas work best with theism, not atheism. Gray tells readers that nature is filled with "unmistakable and irresistible indications of design."

1861   The third edition of *On the Origin of Species* is published.

Henslow's death prompts Darwin to note that Henslow had a bigger influence on his career "than any other . . . I fully believe a better man never walked the earth."

In *On the Relations of Man to the Lower Animals*, Thomas Huxley notes that "the question of questions for mankind—the problem which underlies all others, and is more deeply interesting than any other—is the ascertainment of the place which Man occupies in nature and of his relations to the universe of things."

Archaeopteryx fossils are discovered in Upper Jurassic limestones in Germany. Darwin references the fossils in the later editions of *On the Origin of Species*.

1862   Louis Pasteur, best known today for his contribution to pasteurization, refutes the idea of spontaneous generation.

William Thompson, known as Lord Kelvin, uses a theory of heat flow to estimate that Earth is 20–40 million years old. Kelvin did not know that as Earth cools, new heat is being generated by radioactivity. When the effects of radiation are factored in, scientists calculate that Earth is more than 4 billion years old, a time more than sufficient for biological evolution.

Charles Darwin publishes *On the Various Contrivances by Which British and Foreign Orchids are Fertilized by Insects*.

FitzRoy publishes *Weather Book*, a 400-page book about meteorology based on a recurring mantra: "It should always be remembered that the state of the air *foretells coming* weather, rather than indicates weather that is *present*."

1863   Thomas Huxley's *Evidence as to Man's Place in Nature* argues for the simian ancestry of humans.

Darwin writes to a colleague that the discovery of *Archaeopteryx* "is by far the greatest prodigy of recent time. It is a grand case for me, as no group was so isolated as birds."

1864   Charles Darwin is awarded the Copley Medal by the Royal Society. This medal is the highest honor bestowed by the Royal Society, whose purpose is to glorify God and improve man's estate. Although Darwin did not receive the Copley Medal for his theory of evolution, the Royal Society makes amends after Darwin's death by founding a Darwin Medal whose first three recipients are Wallace, Hooker, and Huxley.

1865   Robert Fitzroy, the *Beagle*'s captain who regretted his role in helping Darwin form his theory of evolution by natural selection, commits suicide.

Gregor Mendel, a 42-year-old monk, presents his ideas about inheritance to a largely uncomprehending Natural Science Society in Brünn (now Brno, in the Czech Republic).

1866   Mendel publishes his ideas about inheritance, which ultimately revolutionize genetics. Two years later, Mendel is elected an abbot and abandons his plant-breeding research. Mendel's work is not widely recognized until it is "rediscovered" in 1900 by Carl Correns, Erich Tschermak, and Hugo Marie de Vries.

Ernst Haeckel meets Charles Darwin at Down House, later saying that "it was as if some exalted sage of Hellenic antiquity . . . stood in the flesh before me." Haeckel's ideas extend far beyond science and embarrass Darwin.

The Geological Society of London awards Lyell its Wollaston Medal, the same award it had given to Darwin 7 years earlier.

Darwin responds to a question about the compatibility of his theory with God by noting that "it has always appeared to me more satisfactory to look at the immense amount of pain and suffering in this world as the inevitable result of the natural sequence of events, i.e., general laws, rather than from the direct intervention of God."

Darwin publishes the fourth edition of *On the Origin of Species*.

1868   Darwin publishes *Variation of Animals and Plants under Domestication*, as well as its reprint.

The American Museum of Natural History is founded.

1869   Darwin publishes the fifth edition of *On the Origin of Species*.

Wallace recants parts of his endorsement of evolution by natural selection while endorsing spiritualism and miracles. This prompts Darwin to send Wallace a letter saying "I hope you have not murdered too completely your own and my child . . . I differ greatly from you, and I feel very sorry for it." Wallace takes a great interest in séances, which Darwin later describes as "rubbish."

Wallace publishes his famous travelogue *The Malay Archipelago*. This book, which Wallace dedicated to Charles Darwin, is still in print.

1871   Darwin publishes *The Descent of Man*, which eliminates the possibility that evolution does not apply to humans.

1872   Darwin publishes the sixth and final edition of *On the Origin of Species*. This edition contains an additional chapter dealing with objections that have been raised about Darwin's theory.

Darwin publishes *The Expression of the Emotions in Man and Animals*.

An attempt to elect Darwin as a Corresponding Member of the Zoology Section of the French Institute fails when the proposal gets only 15 of 48 possible votes.

Darwin publishes the second editions of *The Descent of Man* and *Structure and Distribution of Coral Reefs*.

1873   German Darwinist Ernst Haeckel coins the term *ecology*.

1875   Darwin publishes *Insectivorous Plants* and *Climbing Plants*, and the second edition of *Variation of Animals and Plants under Domestication*.

Lyell dies in London and is buried in Westminster Abbey. The twelfth and final edition of his *Principles of Geology* is published posthumously. The Geological Society begins awarding the Lyell Medal, which features an image of Lyell on one side and the columns of the Temple of Serapis on the other. Darwin does not attend Lyell's funeral.

1876   Asa Gray's essay "Evolutionary Teleology" delineates a middle position in the Darwinian debate.

Wallace's *Geographical Distribution of Animals* establishes modern zoogeography.

Darwin publishes *Effects of Cross and Self-Fertilisation in the Vegetable Kingdom*.

1877   Darwin publishes the second edition of *Fertilisation of Orchids*.

Darwin publishes *Different Forms of Flowers on Plants of the Same Species*.

1879   The eccentric Herbert Spencer, who introduced the phrase "survival of the fittest," publishes *Principles of Ethics*. Spencer's followers later codify his ideas into a school of thought known as Social Darwinism, which argues that the disparity of rich and poor is not an injustice, but simply biology. Industrial pioneer Andrew Carnegie later refers to Spencer as "Master Teacher," and Alfred Russel Wallace names one of his children Herbert Spencer Wallace. Despite his initial popularity, many of Spencer's ideas are rejected before his death in 1903.

Charles Darwin publishes *Life of Erasmus Darwin*, a biography of his famous grandfather.

In his *Autobiography*, Charles Darwin writes that "Considering how fiercely I have been attacked by the orthodoxy, it seems ludicrous that I once intended to be a clergyman," and later noted that "the mystery of the beginnings of all things is insoluble by us; and I for one must be content to remain an agnostic."

1880   Charles Darwin publishes *The Power of Movement in Plants.*

1881   The British Museum of Natural History opens.

Darwin publishes *Formation of Vegetable Mould through the Action of Worms.* This book, like his others, supports Darwin's ideas about evolution.

1882   Charles Robert Darwin dies at Down House and is buried near Sir Isaac Newton and Sir Charles Lyell in Westminster Abbey in London. His tombstone bears the simple inscription: "Charles Robert Darwin, Born 12 February 1809, Died 19 April 1882." The London *Times* notes that "The Abbey needed [Darwin] more than [Darwin] needed the Abbey." To the end of his life, Darwin never publishes anything about religion, noting that what he believed was "of no consequence to anyone but myself."

1883   Francis Galton, Charles Darwin's cousin, outlines the scope of eugenics in his book entitled *Inquiries into Human Faculty and Its Development.*

1884   Gregor Mendel, whose work would later revolutionize genetics, dies in obscurity.

1889   Wallace writes the book *Darwinism.*

Dutch botanist Hugo de Vries' *Intracellular Pangenesis* proposes hereditary factors called *pangens* (sometimes translated into English as *pangenes*). *Pan* was later dropped, leaving the word *gene*, which was first used by Wilhelm Johannsen in 1909.

1891   Eugene Dubois discovers Java Man (*Pithecanthropus* [now *Homo*] *erectus*) in Indonesia.

1892   George Price claims that evolution and socialism are destroying morality, and advocates the biblical flood as the cause of

Earth's geological formations. Price insists that deviations from Ussher's chronology are "the devil's counterfeit" and "theories of Satanic origin."

1893   Wallace is elected a Fellow of the Royal Society.

In his encyclical *Providentissimus Deus* ("On the Study of Holy Scriptures"), Pope Leo XIII defends the inerrancy of the Bible while claiming that "There can never, indeed, be any real discrepancy between the theologian and the physicist, as long as each confines himself within his own lines."

1894   The first skeletal remains of *Homo erectus* are discovered in Java. *Homo erectus* dates back to about 1.5 million years ago.

1895   Thomas Huxley, "Darwin's Bulldog," dies. Huxley's grandsons Julian and Andrew later become famous biologists (with Andrew winning a Nobel Prize in 1963), while their brother Aldous becomes a well-known writer (e.g., *Brave New World*).

Andrew Carnegie endows a museum in Pittsburg whose most famous attraction is a gigantic (90' long) new species of sauropod named *Diplodocus carnegiei*. Carnegie sent full-size replicas of the dinosaur to museums all over the world.

1896   Emma Darwin dies peacefully at age 88 at Down House. Throughout their marriage, Emma—a devout Christian—found Charles' theory to be "painful."

Henri Becquerel heralds the application of physics to measure Earth's age when he discovers the natural radioactive decay of uranium from one of its isotopes to another.

1900   Gregor Mendel's work is "rediscovered" by de Vries, Correns, and Tschermak.

John Scopes is born in Paducah, Kentucky. In 1925, Scopes' trial in Dayton, Tennessee, will become the most famous event in the history of the evolution–creationism controversy.

1906   George Price argues that Darwin's theory is "a most gigantic hoax." Decades later, Price's ideas become the foundation for "creation science."

1907   Indiana becomes the first state in the United States to pass a sterilization law aimed at people in prisons or psychiatric wards who were considered to be "defective." By 1930, 30 other states will adopt similar laws. By 1960, more than 60,000 people in the United States will be forcibly sterilized.

Charles Walcott discovers an assemblage of worms, jellyfish-like animals, and other creatures in the Rocky Mountains of British Columbia, near the Burgess Pass. The so-called "Burgess Shale" contains many animals that arose from unknown ancestors in

Precambrian times, and all became extinct. One of the creatures, having seven tentacles on its back and seven pairs of stilt-like legs, is named *Hallucigenia* because Walcott could not believe what he was seeing.

1908   G. H. Hardy and W. Weinberg develop a mathematical proof showing how genes behave in populations.

1910   The Eugenics Records Office opens as a growing number of people begin to claim that a person's character and abilities are determined at birth by the power of inheritance.

Alfred Russel Wallace receives the Order of Merit.

1910–15   Wealthy patrons of a Bible institute in Los Angeles fund publication of *The Fundamentals: A Testimony to the Truth,* a series of pamphlets reasserting certain truths that had been questioned by scientists and other scholars. *The Fundamentals* are given to every pastor, professor, and theology student in the United States. Although some essays in *The Fundamentals* accepted the idea of evolution, the appearance of *The Fundamentals* is a convenient landmark for the revival of fundamentalism, which has strongly opposed the teaching of evolution.

1912   Geologist Alfred Wegener proposes the theory of continental drift. This theory, which proposes that 200 million years ago the Earth had been one giant continent that broke and drifted apart, helps explain the distribution of organisms among today's land masses.

1913   Alfred Russel Wallace dies at age 90 in Dorset, England. Until his death, Wallace claims that natural selection cannot explain features unique to humans, such as "calculations of numbers" and "ideas of symmetry." As a result of this, as well as his defense of séances and his belief in "the unseen world of spirit," Wallace often puzzles those who admire his other work.

1915   Geneticist Thomas Hunt Morgan produces a new theory of heredity that helps revitalize Darwinism and later becomes a basis for the Modern Synthesis in biology.

A medallion bearing Wallace's name is placed in Westminster Abbey.

1918   President Theodore Roosevelt, who founded America's Natural Parks, writes that he "sat at the feet of Darwin and Huxley" to explain his lifelong interest in evolution and natural history. Many subsequent U.S. Presidents will reject evolution.

1921   Moon's best-selling *Biology for Beginners* textbook states in the preface that biology is "based on the fundamental idea of evolution" and avers that "both man and ape are descended from a common ancestor."

1922	The American Association for the Advancement of Science passes a resolution supporting the teaching of evolution.
1924	John Butler, worried that his children will be corrupted by the teaching of evolution in the public schools, drafts legislation banning the teaching of human evolution in Tennessee's public schools. Seven months later, the Butler Law will be used in Dayton, Tennessee, to prosecute John Scopes in the famous "Monkey Trial."
	South African anatomist Raymond Dart chips the first known skull of an australopithecine child from a slab of limestone.
1925	Tennessee Governor Austin Peay signs John Butler's legislation into law. The Butler Law makes it a crime to teach human evolution in Tennessee and leads to the prosecution of John Scopes in Dayton, Tennessee, 4 months later.
	The spectacular Scopes "Monkey Trial"—one of the original "trials of the century"—takes places in Dayton, Tennessee. Scopes is convicted of the misdemeanor of teaching human evolution and is fined $100 plus court costs. The biology textbook used by Scopes, *A Civic Biology*, reflects the popularity of eugenics-based ideology, noting that "If [criminals and other immoral people] were lower animals, we would probably kill them off to prevent them from spreading. Humanity will not allow this, but we do have the remedy of separating the sexes in asylums or other places and in various ways preventing intermarriage and the possibilities of perpetuating such a low and degenerate race." Leonard Darwin (Charles Darwin's son) sends Scopes a letter of encouragement: "To state that which is true cannot be irreligious ... May the son of Charles Darwin send you in his own name one word of warm encouragement." Following Scopes' conviction, the word *evolution* disappears from biology textbooks.
	Haeckel's version of Darwinism is incorporated into Adolf Hitler's *Mein Kampf*.
1926	Russian Sergei Chetverikov shows that natural populations contain much unseen variation in the form of recessive genes. Later, American geneticists such as Dobzhansky and Wright make similar claims.
1927	John Scopes' conviction for teaching evolution is set aside by the Tennessee Supreme Court.
	In *Buck v. Bell*, the U.S. Supreme Court supports forced sterilization. The usually compassionate Justice Oliver Wendell Holmes writes in the majority opinion that "three generations of imbeciles is enough."

Figure A 3.4 The Scopes Trial occurred in Dayton, Tennessee, in 1925 in the Rhea County Courthouse. Today, the courthouse is a national historic site that houses the Scopes Trial Museum in its basement. (*Randy Moore*)

1930 R. A. Fisher helps reestablish Darwin's theory of evolution by arguing that natural selection controls the direction of evolution by eliminating harmful mutations and perpetuating useful ones. Natural selection can accumulate the effects of otherwise random mutations; if a gene confers an advantage, and increases the rate of reproduction, its frequency in the population increases.

1931 Julian Huxley, the grandson of Darwin's popularizer Thomas Huxley, declares antievolution "dull, but dead."

1932 Sewall Wright describes his "adaptive landscape" metaphor for evolution, with gene frequencies distributed in "peaks" and "valleys." Wright, who points out that not all genetic changes in species are adaptive, shows that natural selection increases fitness.

     The classic Marx Brothers comedy film *Horse Feathers* focuses on a football game between "Darwin College" and "Huxley College."

1937	In his influential book *Genetics and the Origin of Species*, Theodosius Dobzhansky reconciles the fieldwork of naturalists with the mathematical ideas of population geneticists.
1940s	Biologists confirm the importance of gradualism, natural selection, inheritance, paleontology, and population structure in evolution. This merger, based primarily on the work of Thomas Hunt Morgan, George Gaylord Simpson, Ronald Fisher, J. B. S. Haldane, Sewall Wright, Ernst Mayr, George Simpson, and Theodosius Dobzhansky, stimulates discussions that ultimately produce the Modern Synthesis in biology.
1942	Ernst Mayr's masterwork, *Systematics and the Origin of Species*, establishes the evolutionary significance of geographic isolation and small selective advantages.
	Julian Huxley, Thomas Huxley's grandson, publishes *Evolution: The Modern Synthesis*.
1944	George Gaylord Simpson's *Tempo and Mode in Evolution* links population genetics to the fossil record and shows that the fossil record is compatible with Darwin's theory. Macroevolution in the fossil record can be explained by the accumulated effects of short-term evolutionary processes proposed by naturalists and geneticists. In *Tempo and Mode in Evolution*, Simpson introduces the term "quantum evolution," a precursor of the theory of punctuated equilibrium.
	Oswald T. Avery and his colleagues show that DNA is the hereditary molecule.
1946	The Society for the Study of Evolution is founded. It begins to publish its journal, *Evolution*, the next year. Its first editor is Ernst Mayr.
1948	The Communist Party declares that natural selection and Mendelian genetics are erroneous, thereby allowing botanist Trofim Lysenko to introduce his own version of Lamarckism in their place. The results are disastrous for Soviet agriculture.
1950	George Gaylord Simpson makes his famous observation: "Evolution has no purpose; man must supply this for himself."
	Pope Pius XII issues the encyclical *Humani Generis*, which allows that evolution is not necessarily in conflict with Christianity.
	German entomologist Willi Hennig publishes the book that will become (in its 1966 translation) *Phylogenetic Systematics*. He argues that shared, derived characters must be the basis of a classification system that reflects evolution, and that such an analysis must be explicit, allowing for further testing and falsification.

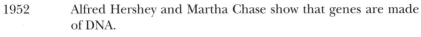

1952	Alfred Hershey and Martha Chase show that genes are made of DNA.
1953	Watson and Crick announce the double-helix structure of DNA.
1958	Biologists learn that DNA controls cellular operations by directing the synthesis of RNA, which in turn directs the synthesis of proteins.
	The Biological Sciences Curriculum Study (BSCS) is founded in Colorado. In the 1960s, biology textbooks prepared by the BSCS emphasize evolution and are very popular. These textbooks "put evolution back in the high school curriculum."
1961	Marshall Nirenberg's discovery that the nucleotide triplet of uracil codes for the amino acid phenylalanine begins the challenge of cracking the genetic code. The code is finally deciphered in 1968.
1962	V. C. Wynne-Edwards, a thoughtful ecologist, attempts to explain reproductive constraint in a "group selectionist" (e.g., "survival of the species") framework.
1964	William Hamilton's theory of kin selection argues that altruism is beneficial to the extent that the recipient shares the genes of the altruist.
	Cambridge opens Darwin College for postgraduate and postdoctoral students. Among its renowned alumnae is Dian Fossey, who interrupted her famous studies of African mountain gorillas to earn a degree there in the 1970s. The College resides in a building that had been the home of Charles Darwin's son George, who was a professor of astronomy at the university.
1966	George Williams responds to group selection arguments with *Adaptation and Natural Selection,* a forceful argument for selection at the level of the individual organism.
1968	Biologists complete deciphering the genetic code. They will complete the Human Genome Project 33 years later.

At this point, we end our timeline—*not* because nothing else has happened—quite the contrary!—but because the study and ongoing discovery of evolution have exploded across our culture and our world. With every new discovery of a fossil and with every new breakthrough in genetics and molecular biology, we understand more about how evolution produced the living world. Evolutionary biologists continue to explore the nooks and crannies of evolutionary theory. Major and minor squabbles erupt—but these are inevitably about *how* evolution happened, not *if* it happened, and are part of the way science is done, by continual examination of evidence, not by fiat. Countless popular books and articles focus on evolutionary discoveries; few 19th century ideas have such a following in the 21st century. Evolution today reaches into every aspect of our lives, from understanding epidemics to the filming of *Jurrasic Park*. Stay tuned! There's more to come!

APPENDIX 4

MEIOSIS AND CROSSING OVER

The key to sexual reproduction is meiosis, which is the production of haploid cells with unpaired chromosomes. Meiosis reduces the number of chromosomes by half. For example, except for eggs and sperm, each cell in the human body contains 23 pairs of chromosomes. Meiosis in humans produces gametes (i.e., eggs and sperm) with 23 chromosomes (one from each pair).

Here's how meiosis occurs. Although most cells have more than one pair of chromosomes, we will—for the sake of simplicity—follow the fate of one pair of chromosomes. In interphase, each cell has one set of alleles from one parent (the dark chromosome), and one from the other (the light chromosome; Figure A4.1). (An *allele* is a particular form of a gene that carries a certain type of information. For example, if a gene coded for hair color, a given allele might specify brown hair.) The chromosomes we show here are condensed so that you can see them more easily; in reality, chromosomes in interphase are more dispersed and not easily visible. The state of having alleles from both parents is called *diploid,* or double. Here we see a homologous pair of chromosomes—that is, a pair in which each chromosome carries the same genes (e.g., for hair color or eye color), although not necessarily the same alleles (e.g., brown hair, blue eyes). During interphase, the DNA of the chromosomes copies itself (i.e., replicates), thereby producing a homologous pair of replicated chromosomes. In other words, each of the original chromosomes has duplicated itself, and both the original and the duplicate are now called *chromatids.* During meiosis, these two sets of twin chromatids are pulled apart. Each set of twin chromatids is now in its own cell. Each of these cells is called "haploid," or single, because although it has two chromatids, it only has genetic material from one parent.

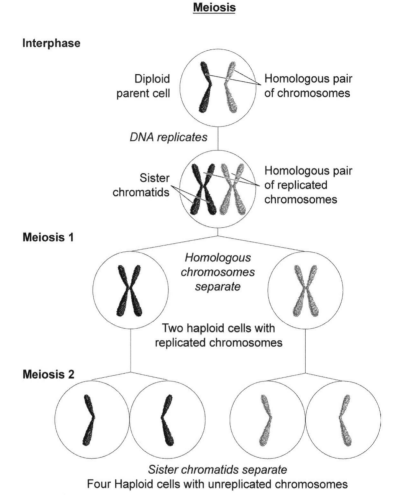

Figure A 4.1 Meiosis. (*Jeff Dixon*)

Meiosis continues into a second stage (called meiosis 2) during which the twin chromatids are pulled apart and sorted into a total of four cells, two cells for each of the twin chromatids. In animals, these cells can become eggs or sperm. When these cells fuse, a diploid offspring is formed.

In the case of crossing over (Figure A4.2), we see the homologous pair of chromosomes after DNA has replicated. Because they are very close together, sometimes parts of chromosomes can trade places—hence the name "crossing over." Meiosis proceeds, but because of the crossing

Crossing Over

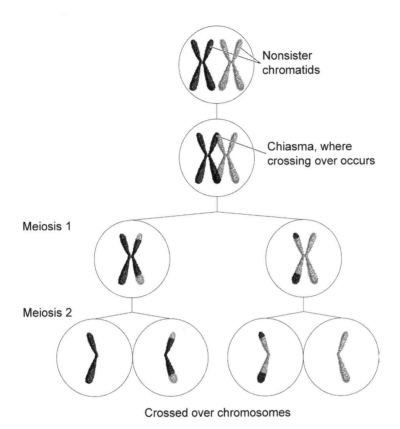

Figure A 4.2 Crossing over. (*Jeff Dixon*)

over, a bit of one parent's genetic material finds itself in a chromosome made up largely of the other parent's genetic material. At the end of meiosis 2, there are four haploid products, as before, but each product differs. This means that a parent is unlikely to produce two gametes that carry the same combination of genetic information.

Although mutations are the ultimate source of new genes, meiosis and crossing over are important because they increase the genetic diversity among offspring. In addition, most organisms have more than 1 pair of chromosomes. If you envision this process happening to several pairs of chromosomes, with each pair sorting independently into haploid cells, you will realize that the potential for diverse gametes—and therefore for genetic variability among offspring—is immense.

APPENDIX 5

THE PRODUCTS
OF EVOLUTION

In this appendix, we will explain in more detail how the most conspicuous residents of the natural world—plants and animals—illustrate evolution by natural selection. In the case of animals, we will show how over a million species of animals are really variations on a few major themes (body plans). In the case of plants, the invasion of land was accompanied by a wide variety of adaptations to the terrestrial environment. After reading this section, we hope that your next visit with nature will remind you of the close relationships we all share.

MANY ANIMALS, FEW BODY PLANS

The diversity of life on earth is truly overwhelming. How could evolution have produced well over a million living, described species, not to mention countless extinct and/or undescribed ones? We have explained the basic mechanisms of evolution in this book. A careful look at the world of animals will reveal that these mechanisms have acted on the variation present within a relatively small amount of fundamental diversity—body plans—to produce vast numbers of species.

There are two great lines of animal evolution based on the number of embryonic tissues. The *Radiata* are diploblastic, meaning that they have two ("diplo") embryonic tissues. The *Bilateria* are triploblastic; they have three ("triplo") embryonic tissues.

The two embryonic tissues of the Radiates cover the outside of the animal (ectoderm) and line the inner cavity (endoderm). A noncellular, jelly-like substance called mesoglea fills the space between the two layers. In contrast, the Bilateria have a cellular layer called mesoderm between the ectoderm and endoderm. The existence of this third cellular layer offers the opportunity for diverse forms within the Bilateria, including a variety of muscles and fancy body cavities.

The Radiates, as their name implies, are radially symmetrical (Figure 4.11). Radial symmetry is circular, much like a pie or a cylinder. There is a top and a bottom; you will get two identical halves no matter how you slice a radially symmetrical object, as long as you slice from top to bottom (or vice versa).

The Radiates can offer a lesson in how many different animals can be derived from one body plan. Consider, for instance, a phylum of organisms called cnidarians. There are over 10,000 described species of cnidarians, which include jellyfish, sea anemones, and corals, as well as the little green *Hydra* that some of you have seen in biology lab. Cnidarians have rudimentary nervous systems and more or less float through life, when they move at all. (Corals, of course, do not change location.) If you look at a typical jellyfish, you will see that its body—mesoglea covered by ectoderm and endoderm—is umbrella-shaped, the rim of the umbrella bearing many tentacles. It floats with its mouth down (more or less), and moves by weak jet propulsion, pushing water through the mouth out of a cavity that is somewhat like a stomach. If you imagine anchoring that jellyfish so that its mouth (along with surrounding tentacles) is pointing up, you have a sea anemone (Figure 4.11). Corals are much like sea anemones that secrete hard, calcium-based exoskeletons. These animals may look superficially different, but all are variations on that jellyfish theme—either floating along, mouth down, or sedentary, mouth up. Natural selection acted on the basic cnidarian body plan to produce an impressive diversity of species, but the fundamental body plan remains the same. (By the way, all cnidarians also share stinging cells called cnidocytes; if you have ever been stung by a jellyfish, you have encountered cnidocytes!)

Meanwhile, you might think that radial symmetry itself is a losing proposition, and this would be understandable, coming from an organism whose ancestors ran, hunted, and evaded predators. What's so great about radial symmetry? Imagine yourself glued to your chair, alone in the middle of a park—*you can't move*. Wouldn't it be nice to have eyes (and ears and arms) all around your body, so you could see (and grab or swat) any approaching food or threat? That's radial symmetry, and it works well for sedentary animals. Most animals, however, move, and that brings us to the Bilateria.

The Bilateria, as their name implies, are bilaterally symmetrical (Figure 4.12). They have a front end, a back end, and two sides. (Think of an earthworm, a cow, or yourself.) There is only one plane in which you can slice a bilaterally symmetrical object and get two identical halves.

The Bilateria have more sophisticated nervous systems and correspondingly more active lifestyles than do the Radiata. After all, if you are moving through the environment, it helps to concentrate sensory function up front—it is better to gather information about where you are going than where you've been. This sensory concentration is called *cephalization* and results in brains, eyes, antennae, and the like in the front end of the Bilateria; their brains are connected to nerve cords that run the length of the animals. In addition, if you are moving through the environment, a streamlined shape is advantageous, especially in the aquatic environments where so much animal evolution occurred. Again, this leads to bilateral symmetry—a long, front-to-back form.

Within the Bilateria there are two great clades: the protostomes and the deuterostomes. These two clades differ in their early development, and form their mesoderm in different ways. The names of the two clades come from the developmental fate of a small opening on the early, multicellular embryo. In the case of the protostomes (*proto* = first; *stoma* = mouth), this opening becomes the mouth. In the case of the deuterostomes (*deutero* = second), this opening becomes the anus, and the mouth develops later. Members of these clades share other developmental traits as well, yet more evidence of shared evolutionary history.

Within the Bilateria, there are numerous examples of body plan modification. We have already mentioned how modifications of the remarkable exoskeleton have allowed over a million species of arthropods to take over earth, air, and sea. Roundworms (also called nematodes, or Phylum Nemata) provide another example. There are over 25,000 described species of roundworms, and they all share similar traits: they are long tubes that taper on each end, covered by a tough cuticle. They all have a tubular gut, with tubular gonads, and they all have four long bands of muscles that help them wriggle along. They live in soil, in water, in hot springs, in Arctic ice, in beer-soaked tavern mats, in plants, and in the intestines of a wide variety of animals. (Even cockroaches have pinworms, very similar to ours!) All 25,000 species (and counting!) of roundworms are variations on that tubular body plan.

The segmented worms (phylum Annelida; 16,000+ species) are yet another example of body plan modification. Whereas the roundworms have a body cavity that extends the length of the worm, the segmented worms subdivide the body cavity, forming a chain of smaller body cavities. In this way, each segment can operate somewhat independently; thus, part of the worm can be long and thin at the same time that other parts are short and fat. This versatility results in remarkable burrowing ability: the worm elongates its anterior (forward) end, with long, thin segments pushing into the soil; they then contract and become fatter, acting as an anchor while the rest of the worm is pulled along. The marine segmented worms probably originated as burrowers, but have diversified. Some live a sedentary life in tubes that they create; they may use their segments to push water, containing oxygen and nutrients, through the tube. Other marine worms move about on the surface and may be active predators. All have the segmented body plan.

Annelids are not strictly marine, however. They have invaded other habitats, including freshwater and land, and this adaptive radiation has resulted in the segmented worms that we call earthworms and leeches. Much like plants, animals also found nonmarine environments to be challenges; in the case of earthworms and leeches, natural selection favored reproductive strategies—egg cocoons—that protected young from drying out or soaking in too much fresh water. Certain segments of the worms were specialized to form these cocoons. Despite these terrestrial adaptations, the segmented body plan, shared with marine relatives, is apparent.

Whereas earthworms reflect their burrowing heritage and consume the soil in which they burrow, leeches are highly specialized to live on fluids that they suck from other organisms. They have two large suckers, one on each end of their bodies, and their internal segmentation is reduced to make room for large digestive pouches that can hold this fluid. Nonetheless, even the highly specialized leeches bear the marks of the annelid body plan—that is, the nervous system, developmental processes, excretory structures, and earthworm-like reproductive structures.

The mollusks are an enormous phylum—almost 100,000 species— and exhibit corresponding diversity. It may be hard to imagine what a clam, an animal that has no brain, might have in common with an octopus, perhaps the most intelligent invertebrate. Once again, the evidence of body plan modification cannot be ignored. This fundamental molluscan body plan includes, among other things, a shell that is secreted by a layer of tissue called a mantle, a muscular foot on which the ancestral mollusk crawled (much like a snail does now), gills suspended in the so-called mantle cavity, the space created where the shell and underlying mantle overhang the foot, and a scraping feeding apparatus called a radula. While the snail crawls across hard surfaces and uses the radula to scrape off algae, clams lead a burrowing existence. The molluscan foot has become wedge-like, and together with the two-part shell, is used to burrow into soft sediment. The head is reduced—almost gone (a common adaptation to a burrowing existence). Siphons funnel water into and out of the mantle cavity, where the gills—greatly enlarged—filter food from the water (Figure 3.3). In contrast, a squid (a close relative of the octopus) is anything but clam-like. It is an active predator, swimming at great speeds. The shell is internal and greatly reduced; the nervous system, so simple in a clam, is complex, with great sensory and mental ability. The mantle cavity has become muscular and when it contracts, it expels water, jet-propelling the torpedo-like squid through the water.

There are several major groups of mollusks, all of which exhibit modifications of the molluscan body plan that have allowed them to invade new habitats and use resources more efficiently. The clam, snail, and squid—burrowing, grazing, and racing through the sea—show us how the same basic ingredients can be modified to produce vastly different, but highly successful, organisms.

The phylum Echinodermata ("spiny skin") contains the sea stars, sea cucumbers, and sea urchins, and shows how a basic five-part body plan can be modified to allow broad habitat and dietary shifts. In addition to their spiny skin, echinoderms have a special kind of plumbing, called a water vascular system, that helps them feed and move. This system reflects the five-part symmetry that characterizes echinoderms and that gives sea stars their five-pointed star shape. Sea stars are predators, and use their rays (and water vascular system) to move and to capture prey.

If you can imagine a sea star that curls its rays up, until the tips meet, and then fills in the gaps, creating a round shell, you have imagined a sea urchin. Again, there is five-part symmetry and a water vascular system, but it is in a different form—wrapping around what is almost a globe—and it is not used for feeding.

Imagine again that you have pulled the five rays of the water vascular system down a long, soft body—that of a sea cucumber. Whereas the sea star and the sea urchin have their mouths directed downward, the sea cucumber lies on its side, using tentacle extension of its water vascular system for picking up deposits and feeding.

The crinoid is yet another kind of echinoderm. Crinoids tend to be sedentary, and their mouth is facing up. Sometimes called "sea lilies," they seem almost flower-like. They use their water vascular system for catching bits of food that descend through the water column. There is a rich fossil record of crinoids, and it is thought that this suspension feeding might be the original use of the water vascular system, with other variations (predation in sea stars, deposit feeding in sea cucumbers, etc.) appearing later.

From sea cucumber to sea lily to sea star, echinoderms used their five-part water vascular system in a variety of ways as they invaded new habitats and found new resources. Despite vastly different lifestyles, they share the five-part body plan, rooted in the water vascular system.

Finally, the chordates—our own phylum—allow us to see the power of body plan modification. Chordates are characterized by the following traits at some stage of development:

Gill slits

Notochord (an elastic rod that provides support for larval or adult chordates)

A postanal tail (i.e., it extends beyond the anus)

Hollow nerve cord

We can identify these traits fairly readily in most vertebrates. In humans, the notochord does not persist into adulthood, but becomes disks sandwiched between the vertebrae (the bones of the backbone). The gills slits are obvious in aquatic vertebrates, but also appear in embryonic terrestrial vertebrates, where they are important in the development of the ear. The dorsal, hollow nerve cord and the postanal tail stay with most vertebrates.

Vertebrates, however, are not the only chordates. There are over 3,000 species of tunicates. These are chordates, but they are sedentary as adults; the larvae are the dispersal forms. The larval tunicate, called a "tadpole larva," has all the traits of a chordate. As it settles and metamorphoses into a sedentary adult, the nerve cord, tail, and notochord shrink. These structures are important in locomotion but less useful for a sedentary existence. The gill slits become larger and larger; they act as a filtering mechanism, and become the feeding apparatus of the adult tunicate.

In the chordates, we have yet another contrast that shows how a fundamental body plan can be modified to work with divergent lifestyles. While all chordates share certain traits, the sessile chordates (tunicates) have emphasized gill slits and use them to filter feed; the nerve cord, notochord, and tail are greatly reduced. In vertebrates, which are far more mobile than adult tunicates, the gill slits are not as noticeable, and the roles of the tail and nerve cord are greatly enhanced.

In summary, the hundreds of thousands of animal species are derived from a handful of body plans—tubes, segmented worms, molluscan shells, and mantle cavities, water vascular systems, and modified exoskeletons, to name a few. Slight variations in these few, basic plans mean that some animals will be able to invade new habitats, use different resources, and provide natural selection with the raw material that powers evolutionary change.

THE EVOLUTION OF PLANTS

Virtually all life on earth depends on plants. Indeed, plants produce the oxygen that we breathe, the wood that we use to build homes, many of the drugs that we use to fight diseases, and the paper that we use to produce books like this one. Without plants, life as we know it would not exist.

The first photosynthetic organisms appeared about 3 billion years ago in the ocean. There, they were buoyed by water, from which they absorbed nutrients. Reproduction was simple—eggs and sperm released into the water met and formed new organisms that floated away, finding new places to live. The first photosynthetic organisms remained in the oceans for 1.5 billion years.

When plants did colonize land, everything changed. Land offered new, unexplored opportunities—for example, more light, nutrients, and CO_2 for photosynthesis—but colonizing land was fraught with problems. The terrestrial environment, unlike the ocean environment, was unpredictable—there was relatively little water, temperatures often fluctuated drastically, and levels of ultraviolet radiation were lethal. The land's air was a powerful desiccant, and the availability of water was unpredictable. As photosynthesis in oceans continued, atmospheric oxygen levels increased and the ozone layer formed, thereby protecting the land environment from most of the damaging ultraviolet radiation.

Colonizing the Land

Colonization of the land began at seashores. There, algae—which are members of a diverse group of organisms called protists—were exposed to freshwater coming from land, as well as an alternating wet and drying environment. Adaptations to survive desiccation would have given these organisms a great selective advantage for colonizing land. Today, most beaches continue to be littered with algae that are washed ashore and can survive periods of desiccation.

Among the first land-dwellers were lichens, which are tough little partnerships between an alga and a fungus; the fungus provides a protective environment for the alga, and the algal cells make food for the fungus. Lichens usually grow in rather inhospitable places—for example, on rocks. They have no roots, and they absorb water directly from rain and air. Lichens grow very slowly.

Plants evolved from a multicellular green alga that lived more than 430 million years ago. As would be expected, plants and green algae share many structural and biochemical similarities; for example, both have the same photosynthetic pigments, (chlorophylls *a* and *b*), both have cell walls that contain cellulose, and both store energy as starch.

Colonization of land was one of the critical events in the history of life on earth, because animals could not colonize land until plants had colonized land. The colonization of land by plants was associated with a variety of evolutionary adaptations:

One of the earliest adaptations to life on land was the *cuticle*, a waxy, waterproof coating of the parts of plants exposed to the drying air. Cuticles protected plants from drying out. Although the cuticle protected plants by keeping water inside the plant, it also kept the CO_2 needed for photosynthesis out of the plant. Plants with small openings called *stomata* that they could open and close to regulate the exchange of oxygen and CO_2 were favored by natural selection.

The earliest land plants needed water for reproduction, because their sperm had to swim through water to fertilize an egg. These plants almost certainly evolved in moist areas. They also formed mutualisms with fungi that could help them procure nutrients in exchange for sugars provided to the fungus by the plant.

Land plants evolved spores and seeds that helped protect their reproductive cells from drying out. Spores and seeds allowed the widespread dispersal of plants. The most successful land plants evolved *seeds*, which are embryos surrounded by nutrients and a protective coat. Because they contain large amounts of stored food and can be dispersed in many different ways, seeds are more effective ways of dispersing plants than spores. When conditions are too wet or dry, or too hot or cold, seeds remain inactive.

The earliest land plants were short and had no true roots, leaves, or stems. Their shortness was necessary: Unlike algae, which were supported by the buoyancy of the sea, the first land plants had to support themselves against the force of gravity. Natural selection favored plants that contained specialized cells and compounds such as lignin, a hard compound that strengthens cell walls and enables plants to support additional weight.

Some land plants evolved vascular tissues specialized to move water and nutrients from one part of the plant to another. Vascular tissue also helped support the plants. Wood, which is formed of several layers of a vascular tissue, enabled some plants to grow to great heights, thereby giving them an advantage in gathering light. Nonwoody plants—also called herbaceous plants—usually have soft, green stems and seldom grow very tall.

The first fossil-evidence of land plants comes from the Ordovician period (510–439 million years ago). During this period, the global climate was relatively mild and shallow, and seas surrounded most continents. DNA-derived evidence suggests that plants colonized land even earlier—about 700 million years ago.

THE MAJOR GROUPS OF PLANTS

Nonvascular Plants

Bryophytes include mosses, liverworts, and hornworts. As their name (i.e., "nonvascular") suggests, bryophytes lack a vascular system. This lack of a vascular system limits their size—most bryophytes are less than 5 cm high, and usually grow as mats close to the ground. Bryophytes do not have leaves, stems, seeds, or roots, and are anchored to the ground by rootlike rhizoids. Bryophytes are often mistaken for algae, with which they share several traits (see above). Like amphibians, bryophytes require liquid water for sexual reproduction; sperm swim through this water to an egg.

There are almost 17,000 different known species of bryophytes, most of which grow in moist environments (e.g., near rivers and streams). Mosses are often the first plants to inhabit a barren area, where they help initiate the development of new biological communities. Peat bogs can cover vast areas, and peat moss (*Sphagnum*) can be mined and dried for use as fuel. Peat moss is often used to pack bulbs and flowers for shipping, and to increase the water-retaining ability of soil.

Seedless Vascular Plants

The seedless vascular plants include whisk ferns (which aren't really ferns), club mosses (which look like miniature pine trees), horsetails (also known as scouring rushes because of their use by American pioneers to scrub pots and pans), and ferns. These plants have vascular tissue for transporting water and nutrients, but do not produce seeds. Like bryophytes, seedless vascular plants require water for sexual reproduction (i.e., so that sperm can swim to eggs). The first seedless vascular plants evolved more than 400 million years ago, and dominated the earth until about 200 million years ago.

Ferns are the most common type of seedless vascular plant. Ferns originated over 350 million years ago and are very diverse—some ferns are less than 1 cm in diameter, whereas others reach heights exceeding 25 m; some ferns are aquatic, and others grow in deserts and above the Arctic Circle. A distinguishing feature of most ferns is the production of spores on the undersides of their fronds (leaves). Their tightly coiled new leaves are called fiddleheads.

Vascular Seed Plants

This group of plants has vascular tissue, true leaves, stems, roots, and seed. A seed is an embryo that is surrounded by nutrients and a tough, waterproof coating. The evolution of seeds was a major advance because it enabled plants to disperse themselves more efficiently. Seeds can remain dormant while waiting for favorable growth conditions. Moreover, when a seed germinates, it contains a store of food for nourishment.

Vascular seed plants do not depend on external water through which sperm swim to an egg. Instead, fertilization occurs after pollen grains (containing sperm) are transferred from one plant to another by wind or by organisms such as insects. The evolution of seeds and the reduced dependence on water are why the vascular seed plants dominate most terrestrial environments.

The two groups of vascular seed plants are distinguished by the type of seed they produce: gymnosperms and angiosperms.

Gymnosperms ("naked seed") produce seeds that are not enclosed in fruits. The first gymnosperms appeared more than 300 million years ago and were the dominant land plants in the Jurassic (208–145 million years ago). There are four groups of gymnosperms:

Conifers, the most familiar division of gymnosperms, include cedar, pine, juniper, spruce, fir, and sequoia. Conifers produce cones, which are specialized reproductive structures composed of hard scales. Conifers' seeds develop in cones, and most conifers have needlelike or scalelike leaves. Conifers are important sources of paper, wood, turpentine, resin, and Christmas trees. Conifers include the world's tallest organisms (*Sequoia sempervirens*, the redwood, whose height can exceed that of a 30-story building) and most massive organisms (*Sequoiadendron giganteum*, the giant sequoia, estimated to weigh more than 6,000 tons). Coniferous forests cover about 20 percent of Earth's land.

Cycads flourished during the age of dinosaurs. These slow-growing plants are native to the tropics and have fern-like leaves that grow atop short, thick trunks. Today, most cycads are grown as ornamental plants.

Ginkgo also flourished during the time of dinosaurs. Only one species (*Ginkgo biloba*) survives today. *Ginkgo biloba* is often called a living fossil because it resembles fossilized ginkgoes that are more than 100 million years old. Unlike most other gymnosperms which are evergreens (i.e., which retain their leaves year-round), *Ginkgo* is deciduous, meaning that it sheds its leaves at the end of the growing season. *Ginkgo*, which can tolerate much air pollution, is an ideal plant for urban areas.

Gnetophytes are a curious group of gymnosperms that have a vascular system similar to that of angiosperms. Examples of gnetophytes include *Ephedra* (the source of ephedrine, a decongestant) and *Welwitschia*, a two-leaved plant only a few centimeters high and more than 3 m wide.

Angiosperms, also known as flowering plants, produce seeds within fruit. Angiosperms include roses, corn, maple, carnations, and wheat. These plants are the most successful plants on earth, and have formed the basis of every major civilization. There are more than 250,000 known species of angiosperms.

Angiosperms are more diverse than gymnosperms; some are less than 1 mm wide, whereas others reach heights exceeding 100 m. Some angiosperms complete their lifecycle in a few weeks, whereas others may live thousands of years. Because of this diversity, angiosperms occupy more niches than gymnosperms. There are two groups of angiosperms: monocots and dicots.

Monocots, which have a one-leafed embryo, include corn and other grasses. Monocots have narrow leaves with parallel veins, and their flower-parts occur in multiples of three.

Dicots, which have a two-leafed embryo, include beans and most trees. Dicots usually have broad leaves with a branching, netlike arrangement of veins, and their flower parts occur in multiples of four and five.

Some angiosperms, such as violets, are herbaceous plants, and others, such as rose bushes, are shrubs. Some angiosperms, such as ivy and grape, are vines, and others, such as oak, maple, and birch, are trees.

Angiosperms appeared in the fossil record about 130 million years ago, and by 90 million years ago were more abundant than gymnosperms. The success of angiosperms is due to a variety of evolutionary adaptations:

The flowers of angiosperms help establish partnerships with animals. For example, flowers' pollen and nectar attract insects, and these insects carry pollen from one plant to another, thereby increasing the odds of cross-fertilization.

Many angiosperms can complete their lifecycles in one growing season. In contrast, many gymnosperms take a decade or more to mature and produce seeds.

The fruits of angiosperms protect and nourish the seeds, and often attract animals that aid in dispersing the seeds.

In warmer environments, the vascular system of angiosperms is more efficient than that of gymnosperms.

Angiosperms are often associated with fungi that help gather nutrients.

Many angiosperms use animals to transport pollen from one flower to another. This means of pollination is more efficient than the wind-pollination that characterizes gymnosperms. However, wind pollination is also used by some angiosperms, such as grasses. In these plants you must look closely to see their small, inconspicuous flowers.

GLOSSARY

adaptation. An inherited trait that improves an organism's chances of survival and reproduction.

adaptive radiation. The diversification of an original species into a group of descendants, each adapted for particular lifestyles.

allele. One of the several forms of the same gene.

altruism. A behavior that benefits others at a cost to the altruist.

Archae. One of two prokaryotic domains of life (the other is bacteria).

artificial selection. The deliberate selection of organisms by humans, performed to emphasize desirable or useful traits.

Beagle. The British ship on which Darwin spent the years 1831–1836. Without his experiences aboard the *Beagle*, it is unlikely that Darwin would have discovered evolution by natural selection.

bilateral symmetry. An arrangement of body parts oriented around a front-to-back axis such that an organism can be divided equally by a single longitudinal cut. A bilaterally symmetrical organism has mirror-image right and left sides.

binomial. A two-part, Latinized name of a species; for example, *Homo sapiens* (humans, or "wise man").

biogeography. The study of the geographic distribution of organisms.

catastrophism. In geology, the claim that most or all geological features are the product of catastrophic events, such as the Noachian flood described in the Bible.

chloroplast. The structure in plant cells in which photosynthesis occurs.

chromosome. A package of DNA and protein.

clade. An ancestral species and its descendants. Clades are based on shared derived traits, which are traits that have appeared relatively recently in an organism's evolutionary history.

coefficient of relatedness (*r*). The percentage of genes shared by two individuals as a result of common descent.

coevolution. Reciprocal adaptation occurring between species.

convergent evolution. The development of similar structures in unrelated organisms as a result of living in similar environments. The development of wings on birds and insects is an example of convergence; in this example, two distinct groups of animals adapted to live in the air.

creationism. The doctrine that each species was created separately in much its present form, by a supernatural creator.

directional selection. A type of selection on a trait that favors individuals with an increased (or decreased) value of the trait in question. Directional selection can result in steady evolutionary change in one direction.

disruptive selection. A type of selection that favors individuals with extreme expressions of a trait and that does not favor organisms having intermediate expression of that trait.

DNA. Deoxyribonucleic acid. DNA stores genetic information.

Down House. The large house outside of London in which Charles Darwin and his family lived for more than 40 years. Today, Down House is a museum open to the public.

eugenics. A social philosophy that advocates the improvement of human hereditary traits through social intervention (e.g., forced sterilization of "unfit" people).

Eukarya. The domain of eukaryotes. Eukaryotes are made of eukaryotic cells, which are cells that contain nuclei and other structures that perform a variety of specialized functions. Eukarya includes all of the protists, plants, fungi, and animals.

evolution. The change in the genetic makeup of a population over time.

exaptation. A trait that may be beneficial, but that had a different function in ancestral species.

extinction. The permanent disappearance of a species or a group of species.

fitness. The relative success of an organism measured by its ability to contribute genes to future generations.

fossil. A preserved remnant or trace of an organism that lived in the past.

founder effect. A change in the gene frequencies caused by the fact that the founders of a new population did not carry all the genes from the source population; a kind of genetic drift.

Galápagos Islands. The volcanic islands 600 miles west of Ecuador visited by Charles Darwin in September 1835. Finches collected at these islands by Darwin and others helped lead Darwin to his theory of evolution by natural selection.

gene. The functional unit of heredity. Genes are specific sequences of DNA that contain information needed to make a protein.

genetic drift. A change in the gene frequencies of a small population due to chance.

genetics. The science of heredity.

Great Chain of Being. A hierarchical system of classification in which organisms have unchangeable positions that reflect their degrees of perfection.

Hamilton's Rule. A rule that predicts when altruism should be favored by natural selection.

homology. A similarity observed in related species that results from their common ancestry.

inheritance of acquired traits. The idea that traits acquired during a parent's lifetime can be passed to offspring; also known as Lamarckian inheritance. This was the basis for the first (albeit inaccurate) theory of evolution.

isolating mechanism. A barrier that slows or prevents gene flow between two populations, thereby allowing new species to form.

kin selection. A type of selection that acts on indirect fitness (i.e., fitness gains from relatives).

Lamarckism. A discredited evolutionary proposal of Jean Baptiste Lamarck claiming that traits developed or lost during life can be passed to offspring.

macroevolution. Major evolutionary changes that involve groups above the level of species.

mass extinction. The collective extinction of large numbers of species in a relatively short period of time. There have been several mass extinctions during Earth's history.

microevolution. Short-term evolutionary change that occurs within a species.

mitochondrion. A structure found in eukaryotic cells which oxidizes carbon compounds and releases energy.

Modern Synthesis. A broad-based effort in the 1930s and 1940s that united the theory of evolution by natural selection with genetics, paleontology, and other disciplines.

mutation. Random change in genetic information. Mutations generate the genetic changes that provide new genetic information.

mutualism. A partnership in which both participants benefit.

natural selection. A difference in survival and reproduction among organisms with different traits.

phylogeny. The evolutionary history of a group of organisms.

population. A group of organisms that can interbreed.

predation. The act of capturing and consuming prey.

punctuated equilibrium. A hypothesis that a species which has been stable for millions of years evolves into new lineages in a period as brief as a few thousand years.

radial symmetry. An arrangement of the body parts of an organism like pieces of a pie around an imaginary central axis. Any cut passing from top to bottom along that axis divides the organism into mirror-image halves.

reproductive isolation. The physical or biological isolation of a group of organisms that prevents them from interbreeding.

sexual selection. Differences in fitness as a result of differences in the ability to obtain mates.

Social Darwinism. A trend in social theory which holds that Darwin's theory of evolution of biological traits in a population by natural selection can also be applied to substantiate a political ideology and critique human social institutions.

speciation. The evolution of new species.

species. The fundamental unit of biological classification. A species is a group of genetically similar organisms that interbreed with each other.

stabilizing selection. A type of selection that favors individuals with inter-mediate forms of a trait and that penalizes organisms with extreme versions of the trait.

symbiosis. Two species living in close association, often physiologically linked.

taxonomy. The science of identifying and classifying organisms according to their evolutionary history.

uniformitarianism. A theory suggesting that Earth's geological features have developed over long periods of time through a variety of slow geologic and geomorphic processes involving common events such as rain, volcanic activity, and wind.

vestigial structure. A structure that has become redundant or useless as a result of evolutionary change.

SELECTED BIBLIOGRAPHY

Barrett, P. H., P. J. Gautrey, S. Herbert, D. Kohn, and S. Smith, eds. *Charles Darwin's Notebooks. 1836–1844*. Ithaca, NY: Cornell University Press, 1987. The text and some interpretation of Darwin's 11 notebooks and related papers. This book provides a behind-the-scenes view of Darwin's thoughts that led to *On the Origin of Species*.

Bowler, P. J. *Charles Darwin: The Man and His Influence*. Cambridge, UK: Blackwell Scientific, 1990. This interesting book focuses on Darwin's scientific ideas.

Browne, J. *Charles Darwin. The Power of Place*. Vol. II of *A Biography*. New York: Alfred A. Knopf, 2002.

———. *Charles Darwin. Voyaging. A Biography*. New York: Alfred A. Knopf, 1995. Browne's two books comprise one of the most comprehensive and detailed biographies ever written. The books discuss virtually all aspects of Darwin's life, ranging from his years aboard the *Beagle* to his family life.

Cox, C. B., and P. Moore. *Biogeography: An Ecological and Evolutionary Approach*. Oxford, UK: Blackwell Scientific, 1993. A short textbook that discusses biogeography.

Darwin, C. *The Descent of Man and Selection in Relation to Sex*. London: John Murray, 1871. In this book, Darwin makes explicit his application of evolution by natural selection to humans.

———. *On the Origin of Species by Means of Natural Selection; Or the Preservation of Favoured Races in the Struggle for Life*. London: John Murray, 1859. This is Darwin's masterpiece. It is also one of the few classic books of science that can be read with pleasure by nonspecialist readers. There were six editions of *On the Origin of Species*, and most scholars agree that the first edition is the most powerful; after you adjust to the Victorian prose, you'll enjoy the detail, thoroughness, and depth of Darwin's arguments. *On the Origin of Species*, which has been reprinted many times in many languages, is an astonishing book.

———. *Journal of Researches into the Geology and Natural History of the Various Countries Visited by the H.M.S. Beagle, under the Command of Captain FitzRoy, R.N. from 1832 to 1836*. London: Henry Colburn, 1839 (2nd ed., 1845). This book, which was published two years after Darwin's return to England, describes Darwin's years aboard the *Beagle*. The book's brisk sales led to a reissue titled *The Voyage of the Beagle*. This book remains a classic among travel-based books.

Darwin, F. *Charles Darwin's Autobiography. With his notes and letters depicting the Growth of the Origin of Species*. New York: H. Schuman, 1950. This, Darwin's most revealing book, was written for his family.

Dawkins, R. *The Blind Watchmaker: Why the Evidence of Evolution Reveals a Universe Without Design*. New York: W. W. Norton, 1986. Dawkins, a brilliant writer, attacks Paley's famous analogy and argues that if natural selection is said to play the role of watchmaker in nature, it is the blind watchmaker.

———. *The Selfish Gene*. Oxford, UK: Oxford, 1976. In this book, Dawkins argues that animals exist for the preservation of genes. This is Dawkin's first and most famous book.

De Beer, G., ed. *Darwin's Notebooks on Transmutation of Species*. Parts I–IV and Addenda and Corrigenda. *Bulletin of the British Museum (Natural History)*, Historical Series, 2(2–6), 1960–1961. An accurate text that helps readers understand how Darwin's ideas led to *On the Origin of Species*.

Dobzhansky, T. *Genetics and the Origin of Species*. New York: Columbia University Press, 1937. In this book, Dobzhansky reconciles the fieldwork of naturalists with the mathematical ideas of population geneticists. This book was one of the driving forces of the Modern Synthesis.

Futuyma, Douglas. *Evolution*. Sunderland, MA: Sinauer Associates, 2005. One of the best modern treatments of evolution.

Gould, S. J. *It's a Wonderful Life: The Burgess Shale and the Nature of History*. New York: W. W. Norton, 1989. This best-selling book describes the evolutionary significance of the strange fossils found in the Burgess Shale. Film buffs will recognize that Gould got the book's title from Frank Capra's movie *It's a Wonderful Life*, which presented the idea of replaying life stories.

———. *The Panda's Thumb: More Reflections in Natural History*. New York: W. W. Norton, 1980. During a period when antievolutionists were reviving the old idea that adaptation implies perfection, Gould used the Panda's "thumb" (it's actually a wristbone) to emphasize the sometimes quirky history of life.

———. *Ever Since Darwin*. New York: Penguin, 1977. A best-selling book about the importance and relevance of evolution.

Hall, B. G. *Phylogenetic Trees Made Simple*, 2nd edn. Sunderland, MA: Sinauer, 2004. This book introduces phylogenetic methods using molecular data.

Herbert, S. *Charles Darwin, Geologist*. Ithaca, NY: Cornell University Press. 2005. A thorough description of Darwin's geological interests and discoveries, and how they relate to his formulation of *On the Origin of Species*.

Keynes, R. *Annie's Box*. London: Fourth Estate, 2002. Keynes, a great-great-grandson of Charles Darwin, describes the personal side of Charles Darwin, focusing on his relationship with his daughter Annie. This is the most emotional and personal of the many books about Darwin.

Lyell, C. *Principles of Geology*. 3 vols. London: John Murray, 1830–1833. Lyell's masterpiece. *Principles of Geology* established geology as a modern science and had a tremendous influence on Charles Darwin.

Malthus, T. R. *An Essay on the Principle of Population, as It Affects the Future Improvement of Society*. London: J. Johnson, 1798. Malthus, a political economist, writes about concerns of the living conditions in 19th century England. He cites three causes (over-reproduction, limited resources, and the irresponsibility of the lower classes) and solutions. Malthus' essay has a tremendous influence on Charles Darwin and Alfred Russel Wallace, the co-discoverers of evolution by natural selection.

Mayr, E. *What Evolution Is.* New York: Basis Books, 2002. This is probably the best available outline of Darwinian evolution. It is written for nonscientists, but it is not overly light reading. However, if you're willing to make the effort, you'll be richly rewarded by reading this book.

Miller, K. *Finding Darwin's God: A Scientist's Search for Common Ground between God and Evolution.* New York: HarperCollins, 1999. Brown, a cell biologist, argues that evolution can be reconciled with religion.

Mindell, D. P. *The Evolving World: Evolution in Everyday Life.* Cambridge, MA: Harvard, 2006. A well-written book about the importance of evolution in our daily lives.

Nichols, P. *Evolution's Captain.* New York: HarperCollins, 2003. A book about the fascinating life and times of Robert FitzRoy, the captain of the *Beagle.*

Pagel, Mark, (editor-in-chief). *Encyclopedia of Evolution.* Oxford: Oxford University Press, 2002. One of the best overall sources of information for all aspects of evolution.

Paley, W. *Natural Theology: Or, Evidences of the Existence and Attributes of the Deity, Collected from the Appearances of Nature.* London: R. Fauldner, 1802. This book presents the classic, and best, argument for natural theology—that is, that one can understand God by studying God's creation. Paley's book, which mentions neither the Bible nor miracles, is the foundation for the "intelligent design" movement.

Palumbi, Stephen R. *The Evolution Explosion.* New York: W. W. Norton & Co., 2001. Palumbi discusses the consequences (e.g., antibiotic resistance) resulting from human activities that have accelerated evolution.

Pennock, R. T. *Tower of Babel: The Evidence Against the New Creationism.* Cambridge, MA: M.I.T. Press, 1999. Pennock explains the nature of science and of evolutionary biology while exposing the fallacies of creationism.

Simpson, G. G. *Tempo and Mode in Evolution.* New York: Columbia University Press, 1944. This book, which is Simpson's seminal contribution to science, integrates paleontology with genetics and natural selection. Simpson, along with Dobzhansky and others, helped produce the Modern Synthesis.

Stanley, S. M. *Earth and Life through Time,* 2nd edn. New York: W. H. Freeman, 1993. This is a thorough introduction to the fossil record and historical geology.

Wilson, E. O. *Sociobiology: The New Synthesis—25th Anniversary Edition.* Cambridge, MA: Harvard University Press, 2000. In this update of his controversial 1975 landmark *Sociobiology,* Wilson argues that social behaviors such as altruism and aggression can be explained by considering the evolutionary advantages of those behaviors.

INDEX

About the Authors

RANDY MOORE is H.T. Morse-Alumni Distinguished Teaching Professor of Biology at the University of Minnesota. He had edited *The American Biology Teacher* and *Journal of College Science Teaching*, and serves on the editorial board of *Journal of Biological Education*. He has won numerous grants and teaching awards, including the Teacher Exemplar Award (Society for College Science Teachers). Moore, an Honorary Member of the National Association of Biology Teachers, has written over 200 articles and books, including numerous textbooks and *Evolution in the Courtroom: A Reference Guide* (2002).

JANICE MOORE is professor of Biology at Colorado State University. She is the author of numerous scholarly articles and the book *Parasites and the Behavior of Animals* (2002). She is also co-editor of *Host-Parasite Evolution* (1997) and serves on the editorial board of *BioScience*.